CITIZENS OF WORLDS

CITIZENS OF WORLDS

Open-Air Toolkits for Environmental Struggle

Jennifer Gabrys

UNIVERSITY OF MINNESOTA PRESS

MINNEAPOLIS · LONDON

The research leading to these results has received funding from the European Research Council under the European Union's Seventh Framework Programme (FP/2007-2013) / ERC Grant Agreement n. 313347, "Citizen Sensing and Environmental Practice: Assessing Participatory Engagements with Environments through Sensor Technologies."

Portions of chapter 2 are adapted from "Citizen Sensing, Air Pollution, and Fracking: From 'Caring about Your Air' to Speculative Practices of Evidencing Harm," *Sociological Review* monograph series 65, no. 2 (2017): 179–92; doi: 10.1177/0081176917710421, originally published by SAGE. Portions of chapter 3 are adapted from "Data Citizens: How to Reinvent Rights," in *Data Politics: Worlds, Subjects, Rights*, edited by Didier Bigo, Engin Isin, and Evelyn Ruppert (Routledge Studies in International Political Sociology, 2019), 248–66. Portions of chapter 4 are adapted from "Sensing Lichens: From Ecological Microcosms to Environmental Subjects," *Third Text* 32, no. 2 (2018): 350–67; doi: 10.1080/09528822.2018.1483884. Portions of chapter 4 are adapted from "Phyto-Sensor Toolkit: Cultivating the Swamps of Urban Air," in *Swamps and the New Imagination*, edited by Nomeda Urbonas, Gediminas Urbonas, and Kristupas Sabolius (London: Sternberg Press, 2022).

Published by the University of Minnesota Press
111 Third Avenue South, Suite 290
Minneapolis, MN 55401-2520
http://www.upress.umn.edu

ISBN 978-1-5179-1404-2 (hc)
ISBN 978-1-5179-1405-9 (pb)

A Cataloging-in-Publication record for this book is available from the Library of Congress.

Available as a Manifold edition at manifold.umn.edu

Printed in the United States of America on acid-free paper

The University of Minnesota is an equal-opportunity educator and employer.

31 30 29 28 27 26 25 24 23 22   10 9 8 7 6 5 4 3 2 1

Contents

Preface and Acknowledgments

Multiple pollutants and substances are churning through the air of modern environments. Toxic gases and intensifying carbon, carcinogenic particles and novel viruses circulate and accumulate into atmospheres that have effects spanning the bodily and the planetary. This book is a study of how people attempt to sense and respond to air pollution. More specifically, it documents and analyzes research from the Citizen Sense project, which worked with communities in the United States and the United Kingdom to monitor air pollution and propose transformations to environments. The Citizen Sense research project, initially funded through the European Research Council, began in 2013 as a study into how sensor technologies organize, promise, and activate citizenly engagements. This research looks at how people are taking up citizen-sensing devices to monitor air, collect data, inform policy, and act on environmental pollution.

Even more than observing citizen-sensing technologies and practices, however, this research has collaborated with communities to document and map pollution, build sensing devices, work through data sets, propose action points, and circulate data stories to influence policy makers and industry. By working with communities to install and test sensors, while monitoring and analyzing data outputs, we investigated how these devices operate and the potential citizenships they would activate. We also studied the varying effects of citizen data as it is collected, circulated, acted upon, rejected, or ignored.

When the Citizen Sense project began, there were few social-research projects that specifically researched and analyzed how community, citizen, or participatory sensing takes shape through these emerging digital technologies. Many of the early studies in this area had been undertaken within the realm of human–computer interaction, where a focus on developing digital devices and systems has been more central; or in creative practice, where speculative arrangements of technology, environments, and social practice spurred different encounters with environments. I detailed some of these projects in my earlier book, *Program Earth*. Yet the research described here seeks to move beyond a central focus on sensor technology to understand in more depth the concrete social, political, and environmental engagements that are made possible or restricted through citizen-sensing practices over time.

The primary way in which this work pursues this area of inquiry is by investigating how citizenships are operationalized along with sensing technologies in the attempt to act on environmental problems. The focus on the citizen is in part a response to one of the most frequent set of questions that surfaced during this research, namely: Who or what is the "citizen" in citizen sensing, what form of political agency is this, and what forms of membership are mobilized here? This is where the notion of "citizens of worlds" comes into play, as the mutual constitution of political subjects and environments. As a concept, "citizens of worlds" signals how diverse political subjects, collectives, and technologies form with and through multiple milieus, and how they undertake practices that express diverse and diverging environmental experiences. This approach expands into an analysis of citizens as more multi-agential, pluralistic, and collective than a seemingly singular unit of abstract citizenship. But it also points to collisions of worlds, where power to make worlds matter is unevenly distributed and wielded. These are conditions that are central to struggles to make more breathable worlds.

Environments are increasingly sites of pollution, extraction, disaster, and development. Technologies and practices for documenting environmental pollution and destruction are not just capturing evidence of these events; they are also devices, operations, and milieus through which citizens and communities materialize and make sense of environmental problems. By documenting practice-based investigations of sensing technologies used to monitor and evidence concrete environmental problems, this study considers the collaborations, conflicts, aspirations, troubleshooting, disappointments, and political change that are forged in specific sensing projects. In doing so, it seeks to identify practices for engaging with environments and working with digital technologies that generate more expansive possibilities for citizens and worlds. How people work with, respond to, care for, shape, fight for, and transform environments informs the political subjects and collectives that materialize. This is also a way of reworking political subjects through their environmental affiliations and attachments—toward citizens of worlds. Different formations of subjects and worlds are constitutive of distinct possibilities for being in and becoming with worlds. Citizens form through worlds, and their coming into being and political capacities register not just in relation to other political subjects but also through the conditions of those worlds.

While this work is practice-based, it is not a project of making sensors for making's sake. Moreover, the research does not stand back and observe participants using sensors, but instead works through a collaborative set of engagements to understand environmental problems and diverse responses to them. This work includes learning about existing monitoring projects while proposing how to develop potential further sensing practices. It involves ongoing attempts to make sense of data and searching for ways to operationalize observations for more breathable worlds. In this process, some contributors to this research have asked to be identified as project contributors and authors of data, while others have asked to remain anonymous, in which case more generalized descriptions of participants are used. Overall, this approach to research involves mutual investigation, shared learning, and respect for the multiple contributions that can be brought to a research project as a democratic and collective endeavor.

My work on sensors has followed a very long trajectory, and it is difficult to identify a starting point for work that has been ongoing for nearly two decades. I began the core part of the research described in this text while at Goldsmiths, University of London in 2013, and completed it while at the University of Cambridge from 2018 on. Thank you to both institutions, including the departments of sociology where I have been based, for providing lively environments in which to undertake this research over a span of more than nine years.

So many colleagues, friends, researchers, and students have contributed in different ways to this research, from providing logistical support, serving on our advisory board, extending publication invitations and contributing to publications, contributing to events and hosting events, arranging exhibitions, sharing ideas, and much more. A very incomplete list includes Anne Alexander, Barbara Neves Alves, Ramon Amaro, Astrid Oberborbeck Andersen, Christian Andersen, Ilia Antenucci, Nishat Awan, Les Back, Karen Bakker, Andrea Ballestero, Richard Balme, Daniel Barber, Ronita Bardhan, Benjamin Barratt, Andrew Barry, Caroline Bassett, Vikki Bell, Laura Beloff, Bernadette Bensaude-Vincent, Michaela Benson, Erich Berger, Armin Beverungen, Didier Bigo, Alan Blackwell, Zach Blas, Dominique Boyer, Rosi Braidotti, Bruce Braun, Roger Burrows, Bram Büscher, Sebastian Büttrich, Baki Cakici, Dominique Cardon, Ele Carpenter, Vivian Chang, Mel Y. Chen, Katie Cohen, Beth Coleman, Rebecca Coleman, Geoff Cox, Endre Dányi, Didier Debaise, Jennifer Deger, Manali Desai, Tridibesh Dey, Mark D'Inverno, Robert Doubleday, Rachel Douglas-Jones, Vera Ehrenstein,

Bianca Elzenbaumer, Ulrike Felt, Laura Forlano, Kim Fortun, Sarah Franklin, Matthew Fuller, Elaine Gan, Emma Garnett, Bill Gaver and the Interaction Research Studio, Natalie Gill, Olga Goriunova, Lisa Gormley, Christelle Gramaglia, Ros Gray, Andrew Grieve, Michael Guggenheim, Miren Guttierez, Muki Haklay, Gay Hawkins, Yasunori Hayashi, Charles Heller, Stefan Helmreich, Steve Hinchliffe, Tom Holert, Erich Hörl, Cymene Howe, Yuk Hui, Helena Hunter, Engin Isin, Anab Jain, Dan Jones, Finn Arne Jørgensen, Melody Jue, Kat Jungnickel, Sarah Kember, Hannah Knox, Olga Kuchinskaya, Brandon Labelle, Brian Larkin, Bruno Latour, Ingmar Lippert, Jamie Lorimer, Yanni Alexander Loukissas, Andres Luque-Ayala, Ruth Machen, Adrian Mackenzie, James Maguire, Alice Mah, Simon Marvin, Cecilia Mascolo, Karen M'Closkey, Joel McKim, Ella McPherson, Ali Meghji, Doreen Mende, Mike Michael, Stefania Milan, Helge Mooshammer, Louis Moreno, Mónica Moreno-Figueroa, Peter Mörtenböck, Debashree Mukherjee, Rahul Mukherjee, Alex Murray-Leslie, Tahani Nadim, Chloe Nast, Joshua Neves, Daniel Neyland, Christian Nold, Susan Owens, Tiffany Page, Weixian Pan, Sylvain Parasie, Luciana Parisi, Lisa Parks, Doina Petrescu, Lorenzo Pezzani, Peggy Pierrot, Søren Pold, Alison Powell, Maria Montero Prieto, Helen Pritchard, Jane Prophet, María Puig de la Bellacasa, Nirmal Puwar, Dennis Quirin, Andrew Ray, Hannah Redler Hawes, Donato Ricci, Wood Roberdeau, Scott Rodgers, Gillian Rose, Ned Rossiter, Evelyn Ruppert, Alder Keleman Saxena, Sven Schade, Lea Schick, Susan Schuppli, Bev Skeggs, Emily Eliza Scott, Shela Sheikh, Brooke Singer, Vicky Singleton, Joe Shaw, Sam Skinner, Johanne Sloan, Eric Snodgrass, Winnie Soon, Michaela Spencer, Will Straw, Lucy Suchman, Tomoko Tamari, Güneş Tavmen, Alex Taylor, Manuel Tironi, Martin Tironi, Nate Tkacz, Anna Tsing, Lynn Turner, Nomeda and Gediminas Urbonas, Jorge Saavedra Utman, Keith VanDerSys, Lucy van de Wiel, Michiel van Oudheusden, Pauline von Hellermann, Rachel Wakefield-Rann, Antonia Walford, Janet Walker, Laurie Waller, Claire Waterton, Eyal Weizman, Jennifer Wenzel, Sarah Whatmore, Ron Williams, Brit Ross Winthereik, Anne-Sophie Witzke, Nicole Wolf, Mark Peter Wright, Kathryn Yusoff, and Matthew Zook. Thanks are also due to the numerous organizers, hosts, institutions, and audiences who listened and responded to presentations of this work, and who were important interlocutors throughout the development of this text.

Multiple communities, collaborators, researchers, technologists, and designers have contributed to the three case studies I discuss in this book. Thanks are due for the fracking-related research undertaken as part of the first case study, "Pollution Sensing," to participating residents in Pennsylvania, including Frank Finan, Rebecca Roter, Meryl Solar, Vera Scroggins, Chuck and Janis Winschuh, Paul Karpich, Barbara Clifford, John Hotvedt, Barbara Scott, Audrey Gozdiskowski, and Alex Lotorto, along with anonymous participants, as well as Citizen

Sense contributors, including researchers Helen Pritchard, Nerea Calvillo, Tom Keene, and Nick Shapiro, and consultants, including illustrator Kelly Finan, data architect Thiam Kok Lau, web designer Raphael Faeh, filmmaker Catherine Pancake, and atmospheric scientist Benjamin Barratt. I am grateful to Illah R. Nourbakhsh and the Create Lab at Carnegie Mellon University for loaning Speck devices for use in the fracking-related research.

For the second, London-based case study, "Urban Sensing," thanks are due to postdoctoral researchers Helen Pritchard and Lara Houston, atmospheric scientists Benjamin Barratt and Khadija Jabeen, graphic designer Sarah Garcin, web designer Raphael Faeh, materials designer Francesca Perona, data architect Lau Thiam Kok, and electronics designer Adrian McEwen for contributing to the collaborative research, design, calibration, and development of the citizen-sensing technologies used in this research. My gratitude also goes to the Southeast London participants and to the organizations that have hosted workshops and events, including Deptford Folk, Deptford Neighbourhood Action, Pepys Estate, Crossfields Estate, Voice for Deptford, APT Gallery, New Cross Gate Trust, Deptford Lounge Library, and New Cross Learning. Along the way, Citizen Sense also benefited from conversations with Councillor Sophie McGeevor and Christopher Howard at Lewisham Council; Barbara Gray, former mayor of Lewisham Council; Rosamund Adoo Kissi-Debrah of the Ella Roberta Family Foundation; and Ronald Bourne of the Lewisham Environmental Justice Alliance; as well as everyone involved in the New Cross and Evelyn Assemblies.

For the third case study, "Wild Sensing," situated in the Square Mile of the City of London, thanks are due to materials designer Francesca Perona and graphic designer Sarah Garcin. Paul McGann of Grow Elephant developed the planters in which the air-quality gardens were installed at the Museum of London. Thanks also to the Phyto-Sensor Toolkit workshop and walk participants, including Sandra Beeson, Natalia Morris, Beth Humphrey, and anonymous participants who contributed ideas for developing the Phyto-Sensor Toolkit. My gratitude goes to the Museum of London for coordinating this project, with support from Lauren Parker, Oliver Whitehead, Tracky Crombie, and Alwyn Collinson; and to the City of London for supporting and contributing to the organization of this project, including Beth Humphrey, Ben Kennedy, Louisa Tan, Asmajan Noori, and Ruth Calderwood. Special thanks to Sarah Hudson for sharing resources on the clean-air gardens developed by residents in the City of London in 2017. The workshop and walk, as well as the planters, were made possible through additional funding from the Museum of London and Low Emission Neighbourhood (LEN) initiative with the City of London. The LEN is funded by the Mayor of London.

For the AirKit proof of concept research, I thank postdoctoral researchers Joanne Armitage and Sachit Mahajan, materials designer Andrea Rinaldi, graphic designer Sarah Garcin, digital designers and technologists Common Knowledge Co-op (including Gemma Copeland, Jan Baykara, Chris Devereux, and Alex Worrad-Andrews), data architect Lau Thiam Kok, and research assistant Verena Eireiner. I am also grateful to the participants and residents in the Forest Hill area of London, including Clean Air SE23, the Dalmain School, and local residents, who contributed to the development and testing of the Dustbox monitoring kit as well as to the collection and analysis of data and communication of results to wider publics and regulators. While she had gone on to work as a lecturer in computing at the time of the AirKit project, Helen Pritchard should be offered a final thanks in triplicate for her epic generosity that has made for ongoing expansive exchanges with Citizen Sense research.

It has been a mammoth effort assembling so much material into one book. Special thanks are due to Melissa Larner for her expert contributions to manuscript preparation and copyediting. At the University of Minnesota Press, Doug Armato, Zenyse Miller, and Anne Carter brought this text to published form, and Danielle Kasprzak contributed to the earlier version of chapter 1 that informed the overall how-to structure for this book. Terence Smyre provided expert support on developing this publication for the Manifold platform, an impressive and engaging experiment in open-access publishing. I am grateful to everyone at the University of Minnesota Press for their ongoing work in making this a truly exceptional and innovative academic press.

The Citizen Sense project and research would not have been possible without the generous and transformative support provided by the European Research Council (ERC). With initial funding through a Starting Grant (313347, 2013–18) and further funding through a Proof of Concept Grant (779921, 2019–20), this work has benefited from sustained and substantial resources from the ERC. Two seed-funded research grants from Goldsmiths, University of London from 2007–8 and 2009–10 allowed me to develop the initial research on citizen sensing that informed my applications to the ERC. Follow-on funding from the University of Cambridge, including through the Economic and Social Research Council Impact Acceleration Account in 2020, has allowed for further development of public engagement and outreach activities. Thank you to these funders and institutions for their contributions, which have been so crucial for undertaking this research.

Dustboxes, a particulate-matter sensor for monitoring air quality developed by Citizen Sense. Photograph by Citizen Sense.

Introduction

ATMOSPHERIC CITIZENS

How to Make Breathable Worlds

I never want less citizenship, I always want more. More different ways of being in relation. And then we struggle it out, because we struggle with the ways in which they're incommensurate.

LAUREN BERLANT, "On Citizenship and Optimism"

How quick, how shallow, how deep, how possible is your breathing right now?

ALEXIS GUMBS, "That Transformative Dark Thing"

An aerial shot of silty terrain fills the screen. The drone video documents a figure in military fatigues, backed by an armored police vehicle and pointing a weapon skyward. A blue projectile leaps into view, knocking the image from its stable frame. As the recording device steadies and pushes onward after the blow, it traces the long tail of a pipeline under construction. Upturned grasses and topsoil, along with trucks and heavy equipment, mark the landscape. The video documents the development of extractive fossil-fuel infrastructure, an event behind barricades that aerial observation brings into view.

The drone that records these images is sensing and monitoring environmental disruption and the destruction of Indigenous lands from the construction of the Dakota Access Pipeline. It gathers soundless video clips that circulate to online platforms and inform ongoing activism and resistance by drawing people to this site, Standing Rock, as part of the NoDAPL movement. Piloted by Water Protectors including Drone2bwild and Digital Smoke Signals, the pilots describe these drone-sensors as an important way to expose the "truth" of extraction and counter government and industry statements. The drones are being "pushed to their limits" in an attempt to develop alternative practices that can contend with—and overturn—extractive technologies that are part of ongoing processes of colonization and environmental destruction.[1] The pilots describe their drones as airborne "protectors" that provide additional power to expose and protest the pipeline construction. Their videos show the growing movement of people who assemble to protect the land, water, and air. They provide a sense of possibility for different and less destructive engagements with environments and technology.[2] Their practices point toward ways of making more breathable worlds, where subjects and environments—people and land—are involved in more reciprocal exchanges and practices of computing otherwise.[3]

From drone monitoring of pipeline construction at Standing Rock in the Sioux (Dahcotah) Nation, to water testing in Flint, Michigan, radiation testing in Fukushima, Japan, deforestation monitoring in Brazil, and air-pollution monitoring in London, a diverse set of DIY, grassroots, and citizen-led practices is materializing to monitor environments. These sensing practices document pollution of air, soil, water, and ecosystems, and they challenge the destruction of environments. Whether monitoring public infrastructure and utilities, contesting extractive industries, or documenting environmental pollutants and biodiversity loss, such practices seek to generate alternative forms of evidence in place of government or industry data. At the same time, these practices express different worlds of experience along with the multiple political subjects and relations that constitute them.

At Standing Rock, protectors, pilots, residents, and activists flew their drones over the pipeline construction to show that unauthorized development in support of extractive industries was occurring. Even as they documented illegal and destructive activity on their lands, drone pilots were told their flights were prohibited.[4] In Flint, residents observed and lived with the effects of polluted water in their homes. They documented and tested their water and worked with university scientists to analyze samples and communicate findings to policy makers, regulators, and the media. While the evidence of severe pollution has drawn international attention, Flint residents continue to have unsafe drinking water.[5] During the 2011 earthquake and tsunami in Japan that destroyed the Fukushima Daiichi Nuclear Power Plant, Geiger counters sold out as residents attempted to obtain information about radiation levels in the area. The group Safecast developed sensors so that participants could monitor environments and exposure, since they did not have access to or did not trust government data and advice about radiation levels.[6] In forests from the Amazon in Brazil to the Carpathian Mountains in Romania, illegal and rampant logging activity continues apace. Networks of remote-sensing technologies and digital reporting systems track deforestation and send text alerts that attempt to halt illegal logging.[7] And in cities around the world, people are using a battery of equipment, including digital sensors, to monitor and mitigate air pollution while managing their exposure to harmful pollution levels.

Citizen-led digital monitoring now extends to a vast array of different environmental concerns. *Citizens of Worlds* investigates how digital sensing technologies transform environmental engagements. This book primarily focuses on how citizens use sensors and DIY electronics to gauge air pollution. From environmental justice groups monitoring petrochemicals in the Imperial Valley of California, to urban residents tracking exposure to air pollution in India and China,[8] there has been a proliferation of citizen-sensing projects focused on air quality. As one of the deadliest forms of environmental pollution, air pollution is a problem primarily caused by fossil-fuel extraction and use, including for transport, construction, buildings, and industry.[9] Air pollution also now occurs at significant levels due to the atmospheric accumulation of fossil fuels leading to climate change, which can contribute to wildfires, particle formation, haze, and smog.[10]

Citizen sensing is a practice formed through struggles to contend with these changing environments. Here, I consider how environmental monitoring technologies that involve low-cost and accessible digital sensors to monitor environments and collect data attempt to challenge and upend existing forms of expertise and ways of addressing environmental problems. These technoscientific engagements remake the usual approaches to environmental action and demand that other experiences—and worlds—be taken into account. In the process, such practices can also undo the designation of the citizen as a normative nation-bound political subject while recasting the affiliations and possibilities of political life.

However, as I will also discuss here, the promissory aspects of these technologies might equally be analyzed as part of a neoliberal sales pitch, where digital technologies are packaged in a glossy veneer of democratic action that does little to shift the entrenched conditions of environmental pollution or social injustice. While citizen-oriented technologies might promise a straightforward realization of positive political change, they rarely yield such effortless or liberatory outcomes when put into practice. On the contrary, citizen-sensing practices produce data sets that governments and experts often view with suspicion. At the same time, digital participation can lead to the proliferation of more (environmentally destructive) digital devices while the conditions of democratic involvement continue to be eroded and social and environmental injustices are amplified.

To account for these variable sensing practices, I engage with citizens neither as universal human actors nor as icons of technological liberation. Instead, I suggest that the "citizens" in "citizen sensing" are politically activated entities that form through worlds of struggle. People monitor environments to address and reduce pollution and related concerns. In this way, sensing citizens become *citizens of worlds*. With this concept, I offer an approach to citizens where different ways of sensing and being affected by environments can activate, reinforce, or transform political subjects and collectives. Citizens require distinct worlds to come into being and to express political affiliations. Worlds are not containers or discrete spheres but rather are constitutive conditions of exchange. Worlds also form as conditions of proliferating citizenships and struggle, as Lauren Berlant notes in the first epigraph to this Introduction. Yet these collective conditions are not only a matter of human affairs but also involve relations that take hold across more-than-humans, technologies, and milieus. To be citizens in the making requires worlds in the making.[11]

This mutual constitution of citizens and worlds unfolds with and through exchanges that I describe as the *breathability of worlds*. "Breathability" indicates not just the ongoing access to actual air to breathe but also how and whether environments, subjects, and relations can be in constructive exchange. Such exchanges involve reciprocity and mutual benefit as part of forming political subjects and worlds. Breathability articulates possibilities for participatory democratic interaction. As Alexis Gumbs notes in the second epigraph, conditions of breathability align with political potential. Rather than indicating an essential biological state, breathability signals situations of differential confinement or flourishing, restriction or expansion that occur in exchange with other entities and milieus.

To be and become citizens of worlds signals the ability to be in constructive exchange with milieus: to observe and contribute, to listen and be heard. Such exchanges allow for the realization of political and environmental relations that extend into the open air of lived experience rather than close in on the airless confines of the universal citizen. With "open air," I refer to the pragmatist proposal to put ideas to the test through practice and to engage with worlds in process.[12] In the open air, the citizen is not an unchangeable entity. Instead, "the subject emerges from the world" and is contingent upon actual occasions and experience.[13] This text delves into the multiple subjects and practices that materialize when sensing air pollution and struggling toward more livable and just environments. Sensing and breathing are ways of constituting these relations. They are practices that indicate how subjects form through exchanges with environments and struggles for breathability.

Citizens of Worlds examines how sensing technologies are deployed, installed, operationalized, and put to work to support concrete struggles over air pollution and related environmental conditions. It analyzes how these practices become legible as citizenly engagements. This book asks: Who or what constitutes a citizen in citizen-sensing practices, and how do sensors activate different citizens in the making? How are possibilities of citizenship formed within and through world-making exchanges? And what are the worlds that citizens would sense, be constituted by, fight for, and struggle to make more breathable? Rather than adopt an approach that univocally argues for the liberatory or condemnatory aspects of citizen-sensing technologies and practices, I instead consider how sensing technologies become caught up in struggles for breathability.

Questions of who or what is a citizen, as well as what is to be sensed and the worlds that are made and sustained through these practices, arise as key problems that unfold in the course of researching, building, installing, and scrutinizing sensing technologies. These questions ask what contributions citizens can make with environmental sensing technologies as well as how these citizens and practices are constituted or disregarded. While citizen sensing can enable specific actions for addressing environmental problems, it can also give rise to further complications for environmental engagement. Not everyone may have the time or resources to undertake environmental monitoring, and different contributions will register more or less forcefully when making claims about environmental pollution. Poor and racialized communities often have less traction when undertaking political advocacy, since environmental claims can be enabled or dismissed based on social, political, and economic privilege. Environmental monitoring can also be a very particular way of configuring environmental problems through data-driven technical practices, which, taken alone, can overlook multiple other experiences. I consider the demonstrations, rejections, and reworkings of citizenship that materialize through citizen-sensing technologies and practices.

In this Introduction, I initially propose *atmospheric citizens* as figures who monitor air pollution as a practice of building more breathable worlds.[14] Yet throughout this study, I also offer multiple other modalities of citizens and citizenship for consideration in the form of instrumental citizens, speculative citizens, data citizens, multiple citizens, and sensing citizens. These are figures constituted through engagement with instruments and toolkits, pollution and harm, evidence and rights, more-than-humans and ecological relations, collective experience and sensation. And yet this is far from a definitive list, since the "citizen" in "citizen sensing" is a shape-shifting entity. In the process of sketching these different yet intersecting citizens of worlds, I consider how subjects embody and express experiences of environmental pollution and destruction. These are citizens in the making, engaged in and constituted through exchanges with worlds. They are figures of breathability and struggle.

In questioning why "citizen" as a term is so frequently appended to digital technologies, this study investigates how such devices on the one hand offer up new participatory potential, and on the other hand create restrictions for democratic engagement.[15] "Citizen sensing" is a term that unevenly describes the use of digital sensors to monitor environments. Indeed, even when working with sensing devices through practices that bear some resemblance to other citizen-sensing projects, Standing Rock drone operators and activists refer to themselves as "protectors," a designation distinct from the language of citizen sensing that expresses ways of watching over and fighting for land. From protectors to community science, to environmental witnessing and citizen science, a diverse array of environmental monitoring practices is now underway.[16] I begin with these examples across multiple sites of environmental struggle because they throw into question how or whether digital sensing technologies and practices might variously be described as "citizen" sensing.

Rather than working with a fixed definition of citizen sensing, I instead consider how environmental monitoring practices and technologies facilitate or hinder more democratic forms of environmental participation.[17] In this sense, I open another line of inquiry distinct from earlier uses of the phrase "citizens as sensors," which described the activities of the "general public" in contributing crowdsourced observations to open-mapping activities.[18] By contrast, I specifically investigate how digital sensing technologies activate more pluralistic practices of environmental citizenship. In other words, I unsettle the figure of the citizen rather than engage with it as a predesignated political actor that might scale into the "general public." At the same time, the technologies and data under investigation here do not assemble into crowdsourced mapping practices. Instead, such an approach considers how the citizenly aspects of sensing do not fully settle in advance of environmental encounters. Sensing, citizens, and worlds differently materialize in ways that can actively constitute or discourage political subjects and relations.

By investigating the problem of the citizen in citizen sensing, this research engages with social, political, environmental, and technological struggles that unfold through diverse monitoring projects and locations. But it primarily focuses on three intensive practice-based and participatory studies undertaken through the Citizen Sense project that I have led since 2013.[19] Our research collective has worked with communities to monitor environmental problems, with an emphasis on air pollution. Drawing on nine years of research, *Citizens of Worlds* documents and analyzes work that has involved developing and testing citizen-sensing technologies, installing sensing kits in collaboration with communities concerned about environmental problems, and analyzing citizen data to generate evidence for action. These three case studies focus on digitally informed ways of sensing air pollution, whether in the gas fields of northeastern Pennsylvania, the congested streets of South East London, or air-pollution gardens in the financial center of London.

Through describing practice-based research developed by the Citizen Sense project working in collaboration with citizens and communities as co-researchers, I investigate how environmental sensing technologies and toolkits take shape in polluted conditions and how struggles arise to fight for more just environments. This practice-based research asks "how to" put sensors to work by undertaking collective research to address environmental problems. In testing these technologies in lived situations, this study documents how these devices work (and fail to function) and engages with the citizen–subjects and worlds within which they become operational. In this way, I engage with citizen sensing less as a topic focused on discrete devices, whether as prototypes or off-the-shelf technologies, and more as formations of political subjects, environmental problems, affected communities, technoscientific practices, political strategies, and worlds struggling to become breathable.

CITIZENS OF WORLDS

In a time when politicians pronounce that "If you believe you are a citizen of the world, you are a citizen of nowhere,"[20] it would seem more important than ever to account for the complex affiliations, attachments, and obligations that form political subjects. Such a declaration seems to announce that citizenship must be singularly designated and tied to a distinct national territory to be rendered meaningful. In its apparent condemnation of elites and jetsetters, this remark entrenches a fixed mode of citizenship with an essential form of belonging. This type of "citizen of the world" has also come under attack by purveyors of nationalism, who insist on discriminatory and racist renderings of the nation-state as a composition of citizens who are seen to "rightfully" belong to its territory.

A citizen of the world might in one way seem to be an elite figure, characterized as much by proselytizing prime ministers as in-flight magazines that promote the benefits of securing multiple national citizenships for weathering global uncertainty. A citizen of the world is critiqued for assuming the privilege of free movement and interchangeable affiliations, along with an uncomplicated appeal to fluid cosmopolitanism. "The world" of which this citizen is a member is a particular designation that can signal privilege undergirded by entrenched inequalities. A citizen of the world might also be diasporic or a subject without a fixed affiliation—yet many migrants find themselves in this situation and are rarely graced with the designation of "citizen of the world." This label is not available to all, however expansive it might seem.

In another way, a citizen of the world might indicate how the "rational demos" is being reconfigured through global interconnection and communications.

This particular articulation of cosmopolitanism designates a world membership that, in its deterritorialization, can be seemingly expansive.[21] A citizen of the world could be someone for whom "local" issues do not define the entirety of their political attentions and engagements, since they are concerned with planetary affairs.[22] Such tendencies become evident within planetary governance, planetary health, and planetary urbanism initiatives. However, the concept of a "world citizen" could also undo the plurality that constitutes the condition of politics by assuming a unitary world as the site of political concern.[23]

Despite these various compositions, the citizen of *the* world would still belong to a "one-world world," as John Law has termed it.[24] The constitution of the citizen and the world—as a citizen of one world—is situated within a universal and undifferentiated rendering of political subjects and the world to which they would belong. The world, in this sense, might even stand in for the singular designation of the nation. Indeed, the citizen of *the* world initially emerged through the transnational flows of colonial trade and conquest.[25] A one-world world can materialize as a figure of domination and extraction, as well as exclusion and marginalization, even as it promises universality.

By recasting the citizen of the world as *citizens of worlds*, this book seeks to study how political affiliations and encounters are multiple and do not necessarily or exclusively parse as nation-state territories or singular forms of belonging.[26] While this is by no means to suggest that struggles for national citizenship are not significant and formidable obstacles for many people, it is also to indicate how citizenships involve multiple exchanges and attachments, in declensions and grammars that differently constitute breathable worlds. There are many other collective entities and identities to which citizens—as variously and unequally constituted political subjects—attach. These are more pluralistic formations of political subjects, which include the nation as just one way of parsing the demos. With this mutual constitution of citizens and worlds, a potential proliferation of citizenships unfolds, forming sites of possibility and struggle because of their plurality and incommensurability.[27]

And yet, "citizen" is likely not even a proper designation of a political subject for all worlds. With this caveat, the term is used here in the plural to signal a differential array of political subjects and environmental relations, as well as a specific entry point for considering how political capacities materialize through digital technologies and practices. The drone pilots and protectors, community scientists, and environmental activists mentioned earlier are political subjects who occupy different configurations of land, collectivity, and more-than-human relations. Rather than innocuously providing data that might facilitate but not challenge standard operating procedures, drone pilots and protectors propel technologies into other encounters that deliberately protest and unearth the violence of a one-world world. Their documentation and witnessing of environmental destruction and extraction might be described as citizen-sensing practices, yet in another way they unsettle the assumed contours of such techno-political undertakings.

The reworkings of citizens and worlds are made through the clashes of settler-colonial states with Indigenous inhabitations, through the protracted battles of residents suffering from environmental racism when living in fence-line communities next to petrochemical industries, by inhabitants dispossessed from their lands due to ongoing and accelerating extractive operations, and by less economically privileged urbanites pushed out of their homes and to the outer edges of cities through forces of development and gentrification. These citizens and these worlds are not the model figures typically imagined when technology companies market drones and sensors and data platforms. Instead, such citizens, formed through struggle, unsettle the seamless narratives of digital participation. In doing so, they demonstrate the limits of these technologies and scripts while forcing different engagements that work toward more breathable worlds.

Proliferating Citizenships

Citizen is a term easily attached to any number of digital technologies. From citizen sensing to digital citizens and internet citizens, numerous digital technologies promise to make us all more informed and active participants. Citizen sensing could suggest the accessibility of these devices to everyday users, or it could signal a frivolous use of the term *citizen* to impart a democratic allure to these technologies. Indeed, at the very moment when digital technology companies are seen to be exercising antidemocratic influences, this packaging of democratic engagement is increasingly used as a strategy, and even smokescreen, to promote an increasing array of supposedly participatory digital products. The "digital citizen" can signal a transformation of political engagement through digital technologies, as well as a possible narrowing of democratic processes through increasing control over data and modes of participation.[28]

Any attempt to locate such a digital citizen within a genealogy inevitably forms a shaky project. The usual designation of the citizen as an ancient Greek conception bound to a city-state, which is now read through the nation, is a way of designating what Engin Isin refers to as an Occidental approach to citizenship. Such an approach to citizenship simultaneously generates specific alterities of citizenship.[29] While it could be possible to analyze the exclusions of citizenship, or who might be designated as noncitizens, Isin's provocation that citizenship is not only Occidental suggests that it might be more productive to consider the alterities that are constituted along with different citizenships, since such alterities do not precede the constitution of specific forms of citizenship. Co-constituted modes of citizenship and noncitizenship are also productive of inequality and struggles for recognition.[30] Such an approach further orients attention to the possible alterities of digital citizenship, where technically oriented forms of political engagement begin to form non- or counter-citizenships to digital citizenships.

Moreover, abstract designations of citizenship manifest differently in everyday practice, and the rights that citizenship would guarantee do not equally extend to all of its members.[31] Writing about the Black Panther Party in the context of the United States in the 1960s and 1970s, Alondra Nelson has discussed how multiple forms of citizenship—across biological, economic, social, political, and other modes of engagement—have not been equally accessible to all people notionally designated as citizens.[32] Referring to this as a "citizenship contradiction," Nelson demonstrates how, enduring the deprivations of these forms of citizenship, Black people sought to expand and claim the full designations of citizenship through political action. They countered these dispossessions through community support programs, health screenings, educational initiatives, and political rallies.[33] Within the context of the civil rights movement, "health rights activism" and radical DIY health initiatives and institutions became a way to "push for equal liberties" and "bridge the stubborn gap that separated civic and social citizenship."[34] As Nelson's work demonstrates, the category of the citizen can be productive of gradated and restricted access to social and political institutions and practices. Struggles for fuller expressions of citizenship often emerge at these junctures, along with attendant anxieties about not being able to inhabit or exercise the modes of citizenship to which one is meant to have access. Yet these struggles can also form other modalities of citizenship that exceed the problem of inclusion in a one-world world to generate other worlds of political possibility.

The practices and proliferations of citizenship, then, destabilize the figure of the universal citizen, demonstrating how it can be discriminatory and exclude other possibilities for political engagement. In this sense, and drawing on Sylvia Wynter's critique of the universal human as an excluding and racializing figure, it might be possible to engage with other and multiple designations of citizenship, especially as projects of citizens and worlds in the making.[35] These subjects further exceed the human in its different declensions to include more-than-human exposures and contributions to the breathability of worlds. Such a move, I suggest, generates citizens of worlds that indicate how other political subjects and exchanges might become possible, especially as they contend with environmental pollution and destruction.

Citizens of worlds is a concept and practice that engages with more than recognition and inclusion within a one-world world. As a concept, it searches for how different ways of being political subjects and making and inhabiting worlds might constitute practices of citizenship. The proliferation of modes of citizenship suggests that there are many ways in which subjects become political. Marisol de la Cadena refers to the multiplications of worlds and ways of being in worlds as the "uncommons," where different worlds exist and come into contact but also diverge and are not always reconcilable.[36] Plurality generates conditions of possibility, yet it is not merely a celebration of the additional. Instead, it can form conditions of struggle within and across multiple worlds that might be incommensurable but can spark encounters and negotiations.[37] Practices that investigate the co-constitutive aspects of citizens and worlds offer a way to recast the hardened origin story of citizenship. They attend to the multiple worldings that generate diverse political subjects and engagements.

The "citizen" as it is operationalized through *citizen*-sensing technologies could at first be a seemingly universal subject and condition. But the expressive political capacities that such technologies are meant to enable do not so easily or evenly confer citizen-like status on everyone, where people with less economic and cultural capital, racialized communities, women, and many others outside the arenas of power and expertise will find that their contributions are less audible or delegitimated within arenas of evidence-making. Instead, questions arise about the proliferating political subjects and relations that take shape along with these practices and technologies. Citizen sensing can be a practice to mobilize findings from citizen data, appeal to policy makers, hold polluters to account, address environmental problems, and make breathable worlds by computing otherwise. Yet these practices do not follow effortless or straightforward trajectories. Regulators often ignore citizen data collected with sensors. "Facts" about environmental pollution can be dismissed if they do not align with sedimented relations of power and privilege. Moreover, the uptake and use of sensors might not always follow the same protocols or patterns of use and observation—not because these are erroneous practices but because they might tune in to different registers of experience and account for other worlds in the making.[38]

Citizens of Worlds examines how such modes of citizenship are constituted along with or even against citizen-sensing technologies as they are used in practice. This approach puts the citizenly aspects of these technologies to work to query the concrete political engagements that occur. Digital devices, in this sense, do not merely enable alternative forms of political engagement, where the citizen and the collective to which it belongs are wired up but remain relatively unchanged. Instead, the conditions for being and becoming citizens, for sensing environments and making evidence claims, can transform and generate altered possibilities for breathability through these technoscientific reconfigurations.

Through grappling with the formation and activation of the "citizen," this research commits to an investigation of the political subjects, relations, and worlds that these technologies generate. Rather than dismissing or discarding the term "citizen" as overly contentious or loaded, and opting for a term such as "community," "civic," "participatory," or another seemingly less charged phrase,[39] I work with this complicated term exactly because it raises questions about the democratic dilemmas and potential of digital participation and environmental action. Once deployed, the term "citizen" opens up many unforeseen detours, obstacles, opportunities, and necessary reworkings in the course of its implementation. Part of the impetus for attending to the "citizen" is to demonstrate how this figure is not, as is customarily assumed in the context of citizen sensing, one that simply expresses the "general public" or an amateur participant who is meant to operate in contrast or in complement to expert science. Instead, the "citizen" in citizen sensing can become an indeterminate entity that forms through struggles toward more breathable worlds.

Although this study undertakes an intensive discussion of citizen-sensing practices and technologies, it also queries and reworks the designations of citizenship and approaches to political engagement that might be mobilized through these practices. Rather than signal toward more abstract designations of citizenship, whether in relation to cities, nations, globes, or planetary governance, *Citizens of Worlds* works through the practices by which world-making and world-binding activities such as environmental sensing also become citizen-making and citizen-binding practices.[40] The conditions of stressed environments, of having to breathe polluted air, and of not being able to alter states of uninhabitability can feel more constricting than expansive, where political subjects are bound to problems with which they are forced to grapple because they affect the very conditions of their breathability. Distinct citizenships materialize through struggles that unfold within these stifling atmospheres.

ATMOSPHERIC CITIZENS: SENSING AIR POLLUTION

The World Health Organization (WHO) has deemed air pollution "the largest environmental risk factor" on the planet.[41] As many as 8.8 million people worldwide die each year from the effects of indoor and outdoor air pollution, with 4.2 million of these deaths attributable to outdoor air pollution.[42] On an annual basis, as many as 800,000 people in Europe and 40,000 people in the UK experience premature death from air-pollution-related causes, with over 9,000 UK deaths located in London.[43] Overall, air pollution causes one in nine of total global deaths. Nitrogen oxides, ozone, volatile organic compounds, particulate matter, sulfur dioxide, and many other pollutants circulate through environments and bodies, contributing to disease and death. Cities from Beijing to Tehran and from London to Los Angeles are blighted by poor air quality. Yet as a global problem, air pollution is differently experienced, monitored, evidenced, and acted upon across the diverse locations that it affects.

The differential conditions of air quality and pollution across disparate locations can indicate the many and intersecting environmental problems, from resource extraction to extensive construction and development, traffic congestion, and petrochemical industries. Among the numerous articulations and proliferations of citizens and citizenship that I develop in this study, *atmospheric citizens* is a specific configuration that signals how air pollution and struggles for breathability affect people. Atmospheric citizenship materializes through the ongoing and worsening problem of air pollution. It designates political subjects and environmental actions that might reconfigure and transform atmospheres.

While this study engages with a limited cross section of practice-based citizen-sensing projects undertaken in the UK and United States that I describe in the chapters that follow, such projects are underway in numerous places worldwide. The Citizen Sense project has discussed monitoring practices with researchers, regulators, and community groups in locations from Vietnam and India to Kazakhstan and Montenegro, and from Chile to France and California. Although similar technologies might be used in many of these sites, very different considerations about atmospheric forms of citizenship are often at play. Issues arise related to the legality of collecting data, the availability or absence of state funding and support, the air-quality indices used, the local weather conditions, the receptivity of regulators to citizen data, and the communities of support both in environmental and technical contexts, which might also be able to act on findings from citizen monitoring.

Citizen-sensing practices and technologies could seem to outline a straightforward way to document, communicate, and act on the problem of environmental pollution and destruction and for individuals to avoid exposure to pollution by monitoring their everyday air space. Yet practices for monitoring air pollution also show how atmospheres are unevenly experienced, sensed, and acted upon through bodies, sensors, and environments. The atmospheric exchanges that sensors and sensing practices mobilize then inform the conditions and possibilities of citizenship. These are, in other words, atmospheric modes of citizenship.

When developing a concept of atmospheric citizenship in the context of air pollution, I draw in part on Berlant's notion of "ambient citizenship," where, as they suggest, political world-building projects first become perceptible as atmospheres.[44] It is "the ordinary affective or interactive aspects of social exchange" that make up the "scenes of *substantive* citizenship," even though rational, communicative, and legal registers of citizenship are often (over-)emphasized as the key registers of citizenship.[45] Questions of who takes up space, of whose voice dominates, and how and why, are atmospheric (or in other words, affective) matters in Berlant's rendering of ambient citizenship.

I engage with multiple works that signal the affective and political registers of atmospheres and how they are constitutive and expressive of citizenship, along with research that emphasizes the unevenness and disparity of the atmospheres in and through which citizenship forms, especially in relation to air pollution. These works draw attention to the plurality of atmospheres and exposure to air pollution by capturing different struggles to breathe, which are as much sociopolitical and environmental as they are bodily. The "fact" of needing to breathe cannot be described simply as a universal condition when lived atmospheric conditions vary so significantly. Instead, such atmospheric conditions require grappling with the everyday and infrastructural conditions of environmental violence that constrain the ability to breathe.[46] Atmospheric citizens form as subjects and environments, or in other words, as citizens of worlds informed by the constitutive aspects of breathing.

Combat Breathing

Atmospheres are expressive of the inhalations and exhalations of everyday life. Frantz Fanon elaborated on this condition of atmospheres through his investigations into colonial violence. Writing in the context of colonial occupation in Algeria, he argues that colonialism is not only an "occupation of territory." Instead, colonial occupation extends to a country's "daily pulsation." Within this pulsation, individuals undertake "occupied breathing" that Fanon suggests can become a form of "combat breathing," as it simultaneously endures yet works against the occupation of daily pulsations.[47] Combat breathing is a mode of respiration that contests its own occupation and suffocation.[48] Fighting for breath could on one level be a practice of fighting for survival. Yet on another level it could also involve fighting for different relations and inhabitations that are not bound to colonial power dynamics infusing everyday life. Fighting for breath consists in fighting for worlds.[49] Rather than referring to a more universal or biological rendering of breath, combat breathing marks out a struggle to transform the specific occupied atmospheres of everyday life and, in so doing, to cultivate more breathable worlds. Less an absolute envelope or sphere that conditions and terminates breathing,[50] such an approach draws attention to modes of exchange as well as possibilities for breathing otherwise. Within this context, atmospheric citizens materialize as political subjects who come into being as they struggle toward the decolonization not only of land but also of everyday pulsations and exchanges.

Combat breathing is a practice and analytic that connects struggles across bodies, politics, histories, and environments.[51] Writers and theorists from Christina Sharpe to Alexis Gumbs have taken up Fanon's discussion of combat breathing to discuss on the one hand how toxic atmospheres become a sort of "weather" in which Black people struggle to breathe, and on the other hand to convey the violence of being robbed of breath within actual conditions of pollution, assault, and deprivation. In explicating Fanon, Sharpe suggests that it is necessary to turn to "the totality of the environments in which we struggle; the machines in which we live" to grapple with the "weather" of un/breathability.[52] Indeed, Sharpe proposes strategies for cultivating breathability—or breathing otherwise—by "refusing nation, country, citizenship" as anti-Black formations that contribute to unbreathability.[53]

In their discussions of breathability, both Gumbs and Sharpe refer to the well-known words of Eric Garner, who, when being assaulted in 2014 by NYPD police officers, repeated eleven times, "I can't breathe." His words, and his death from this restraint of breath, have become a central reference point within the Black Lives Matter movement. The repeated enunciation of "I can't breathe" by activists struggling for social justice recalls the violent death of Garner and many others. It also calls out the confined, airless, and toxic atmospheres within which Black people find themselves struggling to breathe due to systemic racism. This phrase gained renewed relevance in 2020 after the murder of George Floyd, whom a Minneapolis police officer suffocated with a knee on his neck as Floyd repeated, "I can't breathe." This call to breathability has resounded throughout protests in the United States and cities worldwide, as struggles for racial, social, and environmental justice amplify and gather force.

Writing in an earlier context, Fanon noted that revolutions emerge—here describing Indo-Chinese people rising up against French colonialism—"because 'quite simply' it was, in more than one way, becoming impossible for [them] to breathe."[54] Pheng Cheah refers to Fanon's discussion of revolution and struggles to breathe to show how decolonial efforts can lead to the formation of new subjects, along with new worlds in which they can breathe.[55] Struggles to breathe are articulations of other ways to move and respire that demand an expansion of sociopolitical possibilities. These are struggles with and against power, inequality, and the diminishment of worlds that people inhabit, require, and seek to build. Such an approach diverges from understanding atmospheres and breathing as universal components of life to demonstrate how atmospheres and breathing are formative and transformative in addressing sociopolitical, epistemic, and ontological injustices.[56]

The relationship between air pollution and violence, particulates and power, as Lindsey Dillon and Julie Sze point out, leads to conditions of "embodied insecurity through the everyday act of breathing," especially for racialized and low-income communities. Not only does the phrase "I can't breathe" signal this insecurity, but it also indicates the "racial health disparities" and environmental injustices that lead to higher rates of asthma for Black people in the United States and elsewhere.[57] Environmental injustice in the form of air pollution occupies and constricts breathing. Such inequalities sediment in environments, bodies, and relations, which people struggle to transform by making more breathable worlds.

The phrase "I can't breathe" is a call to breathability, which, as Gumbs writes, is "designed to help us remember how to breathe and how to invite our revolutionary ancestors into our bodies and our movement."[58] From these "how-to" practices for breathing, Gumbs outlines "a cosmology" that forms to connect multiple struggles for justice.[59] The respiratory and inspiratory exchanges taking place here are a generative and revolutionary mixing of earthly relations and struggles, inheritances and embodiments, which work against conditions of unbreathability. Gumbs develops combat breathing toward "Black feminist breathing," as a practice that attends to "a lineage of Black revolutionaries whose faith in freedom continues to inspire."[60] Inspire and respire are exchanges informed by combat breathing, where to carry on breathing is to find ways to make that breathing less onerous and more expansive. Combat breathing troubles the divide between respiration and inspiration, not as a blindly hopeful project but as one that reckons with injustices while refusing to be bowed down, drowned out, or suffocated by them.[61]

Breathing Collectives, Breathing Otherwise

Breath is a topic that has received attention from multiple fields. From environmental justice scholars and practitioners outlining long-standing work on the impairments to breathing in the form of asthma, heart, and pulmonary conditions for racialized communities and those who are less economically privileged, through social studies of science and technology that investigate the expert devices and practices that test and regulate the conditions of breath, to Black studies scholars who scrutinize the colonial legacies of breathing to work toward less suffocating sociopolitical conditions, there is a rich if at times diverging set of analyses on this topic. I especially engage with literature that attends to how atmospheres and subjects, citizens and worlds, are mutually constituted through struggles for justice. I build on and extend this work to consider how citizens are not simply in worlds but rather how they are constituted through exchanges with worlds that express and materialize differential sociopolitical conditions of breathability.

As a process of daily pulsations, breathing is a mode of subjectification that is informed by environments and possibilities for collective engagement across registers of atmospheric exchange. In a similar way, Indigenous literature discusses how citizenship and breath are coextensive, where breath is an articulation of what connects people, land, and organizations in mutual exchange.[62] To be without breath is also to be without freedom, liberty, sovereignty, or citizenship. Breath in this way is not just an exchange; it is also a form of mutual benefit and governance. Writing about Canada's First Nations, Leanne Betasamosake Simpson describes Nishnaabeg "governance as breathing—a rhythm of contraction and release."[63] Such approaches to atmospheric governance are inclined toward reciprocity and flourishing that form through a mutual politics of breath. The daily pulsations identified by Fanon take on another register here, where governance requires sustaining and cultivating collective breathing by passing through combat into worlds of exchange.

Breathing involves more than one person inhaling and exhaling. It involves environments and other entities as they respire and exchange atmospheric gases and pollutants, along with circulations of air and weather. Breathing is transformative, remaking bodies and environments through continual exchanges of substances that accumulate and sediment into new ecologies. Étienne Balibar has suggested that a citizen is necessarily constituted as a member of a political community along with "fellow" citizens and is not a solitary entity.[64] A citizen always belongs to a collective that is the site and process of the political. But "fellow" citizens in the context of breathing, and sensing, extend not just to other humans but also to other entities, environments, and atmospheres that are in process, informing the possibilities for citizens and collectives to take hold as a democratic project. Indeed, breathability undertaken by humans depends upon multiple entities as they constitute and collectively create breathable worlds.

I have previously written about the co-constitution of subjects and worlds in *Program Earth*, which looked more broadly at the proliferation of environmental sensing technologies. I fold in this earlier work here to investigate how breathability is a process that forms subjects and environments through distinct conditions of exchange.[65] Breathing, like sensing, is an exchange that constitutes subjects and milieus, that establishes the ongoing relations that continue to sediment into worlds of experience. I return to these world- and subject-forming conditions to consider in more detail how breathing and breathability are not universal properties of bodies but rather practices with differential possibilities and effects that inform the ability to be and become citizens of worlds. *Citizens of Worlds* expands on this earlier research by investigating how citizens, as political subjects, materialize across human and more-than-human relations and worlds of experience. Breathability signals exchanges with environments and other entities. It is an expression of collective and changing experiences. Incorporating more-than-humans into registers of citizenship necessarily expands the breathability of those worlds while also informing the conditions for being and becoming citizens of worlds.[66]

As citizens of worlds, atmospheric citizens are therefore constituted through distributed and social conditions that make breathing possible. The sociality of breathing and air, as Ashon T. Crawley notes, can generate different currents of air that involve mixing and exchanging as well as openness. Through these dynamics he refers to "otherwise air," which searches toward other possibilities for breathing.[67] Such characterizations of air, atmospheres, and breathing engage with currents and worlds of exchange, transformation, and struggle.

I consider how struggles for breathability surface through environmental sensing practices for monitoring air quality. This research analyzes how such practices could transform, reinscribe, or fail to address the daily pulsations of environmental pollution, injustice, constriction, and violence. Through these practices, I suggest that atmospheric citizens and citizenships materialize that are differently shaped by struggles to build more breathable worlds. To experience the constriction of breath and the pollution of air is to experience the world-binding conditions of everyday life. But citizens and worlds in the making also materialize through struggling against these conditions of unbreathability.

HOW TO MAKE BREATHABLE WORLDS

This investigation into atmospheric citizens, along with the plural figures of citizens discussed in the pages that follow, unfolds through a practice-based and collaborative investigation into how people put sensing technologies to work to make more breathable worlds. At stake here are not just questions of what counts as a breathable world and for whom, but also how these worlds and practices of citizenship can be mobilized. The project and question of "how to" is central to this study, since it allows for an exploration of the practices that guide citizen-sensing efforts. Sounding a pragmatist note, Law suggests that practices are central to how different worlds are made and sustained and that they inform how "different realities are enacted."[68] Even more than *enacting* realities, however, this study investigates how sense-making practices constitute ways of *struggling* for realities that could constitute breathable worlds.

This engagement with practice is at least twofold. It involves not only studying how citizen-sensing practices and technologies address environmental problems but also working with communities to develop practices and toolkits that respond to their struggles for breathable worlds. A study that is both *of and through* practices, this approach turns toward concrete effects to understand how citizen-sensing practices for monitoring air pollution materialize. Practice is a way to undertake research in the making, rather than work from a predetermined position. Practice as research is not, however, "applied" in the sense of a functional exhibition of theory. Instead, it is a way to mobilize and test propositions through actions within distinct conditions and communities of inquiry.[69] The modalities of practice developed here are more resonant with the notion of praxis than making for making's sake, since they also activate and connect with the ongoing formation of political concepts and actions.[70]

As a practice-based study on the plurality of world-making and citizen-sensing practices, this text spans multiple fields, including digital social research and science and technology studies, pragmatism and social theory, Black studies and Indigenous studies, political ecology and environmental justice. Through this approach, and in conversation with these fields, the notion that a device might embody and enable particular forms of citizenship can be tested, challenged, and rerouted. By working with citizen-sensing technologies to question how they allow—or do not allow—for different expressions of citizenship, participants could challenge the claims made about devices while also orienting citizen sensing practices toward more livable and breathable worlds. Practices of citizenship potentially materialize here less as a scripted technological program or sales pitch and more as a contingent and inventive set of collective engagements in the open air. Such engagements are as likely to arrive at impasses and confrontations as they are to realize improved air quality. I highlight the unevenness of these engagements as an indication of how atmospheric and multiple other citizens form and operate.

For each of the three case studies that inform this research, a survey of existing citizen-sensing practices underway in distinct locations became the initial spark for forming collaborations with community groups, individuals, and organizations already involved with monitoring environments. Through fieldwork, interviews, and ongoing documentation, the Citizen Sense research group learned about a wide range of ongoing sensing practices, which became the basis for making new sensing kits with communities. We also continued ongoing conversations with residents and communities, held workshops and hosted monitoring walks, undertook site visits and installed sensors, diagnosed and repaired devices as they misfired and broke down, liaised with regulators and policy makers about environmental monitoring, joined teleconferences and meetings to discuss citizen data, retrieved devices at the end of monitoring tests, analyzed data and built data-analysis infrastructure, wrote and co-wrote data stories, and communicated and circulated findings from citizen data to a wide range of groups, including governmental agencies, the media, and scientists.

While one way to describe this research might be through the lens of participation, I deliberately work with the terms "collaboration" and "struggle," since these concepts emphasize how these projects took shape, often with considerable collective effort.[71] Citizens struggle to make their voices heard when ways of life are at stake due to environmental destruction. Collaborative and community-oriented research can be a site of struggle, as Linda Tuhiwai Smith has suggested. By working through more accountable methods, it might be possible to shift the gaze of research and contribute to the self-determination of communities involved in research.[72] Our collective practices of sensing air pollution were neither a straightforward project of user testing focused on the technical capacities of devices nor a tick-box exercise of gathering public opinion or input. Instead, the citizen-sensing investigations we developed were engaged with ongoing social, political, and environmental problems that closely informed our attempts to research sensors in practice.

As collaborative undertakings, these projects frequently involved longer time frames of becoming familiar with communities' existing monitoring practices and environmental concerns, finding workable practices for coming together to create a monitoring kit, and having multiple meetings and discussions to understand how best to analyze and communicate findings from citizen data. While our research group assembles monitoring toolkits for adaptation and use, the process of building a monitoring infrastructure takes place with communities and in response to their specific concerns. The material gathered here describes and analyzes the complexities of undertaking this practice-based work while continuing to address the key questions of who or what is a citizen and how worlds are formed or sustained through these sensing practices. In focusing on the air and air pollution as a growing area of concern for many urban dwellers, the text examines how the experience and evidence of air pollution contribute to particular ways of organizing environmental struggle and environmental citizenship through lived experiences of breathing polluted air, contending with urban traffic, and enduring ongoing construction and development.

The citizen-sensing practices and technologies discussed throughout this study take the form of open-air toolkits, since on the one hand they deal with matters of sensing air and air pollution, and on the other hand—drawing on and cre-atively extending William James—they are formed through concrete experience and putting ideas and things to work in worlds. These toolkits are in process, gathering force, or dissipating as they are set into practice in the open air to form worlds in a "multiverse."[73] Such an approach allows for an understanding of how open-ended practices settle into recognizable forms as well as how the open, indeterminate, experimental, and speculative aspects of technologies unfold. The openness of technology might be pursued less as a question of open hardware or software and more as an investigation into how open technological engage-ments might be rerouted to be more democratic and inventive. These lines of inquiry require attention to and engagement with digital technologies as they are taken up, used, and reworked through practice.[74] Openness operates here in a pragmatist register, where how instruments are put to work in the open air informs the subjects and milieus that take hold.

Practice-based research demonstrates how open-ended and inventive encoun-ters with digital technologies might be one way of more fully researching and addressing the qualities of technological engagements. They might also be ways to work toward more equitable and less extractive technologies and technological practices, when discrimination and inequality can unfurl through the very "code" of these devices.[75] With these points in mind, citizen-sensing research and prac-tice might expand from the usual framing as sensing technologies enabling the collection of monitoring data toward political action to encompass a more inven-tive and open set of engagements. Digital technologies could be encountered as always in the making, changing through practice, and also open to disruption through sites of active engagement.

How to Use This Book: A Chapter Guide

Citizens of Worlds takes the form of a how-to guide of sorts that presents the practice-based research of making toolkits and working with communities to sense air pollution. Each chapter examines a mode or practice of citizens and citizenship. Each chapter also explores different modalities of the "how-to" to analyze how practices, citizens, and worlds materialize. How-to guides are now proliferating along with any number of DIY engagements, from kits for build-ing sensors, guides for launching satellites, instructions for managing urban infrastructure, and campaigns for achieving political change. This study takes seriously the upsurge of the how-to guide as a literary genre and social move-ment that attempts to give voice and direction to political and environmental struggles. At the same time, the research works through the opportunities and limitations of the how-to guide in providing apparently clear instructions on how to address a world—or worlds—gone wrong. This investigation into how to make breathable worlds involves attending to citizens and worlds in the making. "How-to" involves putting technologies and practices to work, understanding their effects, and transforming conditions toward more breathable worlds.

The citizens, worlds, and how-to practices in the making that these chapters describe are far from definitive and could proliferate as an endless list. Chapter 1, "Instrumental Citizens: How to Retool Action," is an unconventional methods chapter that examines more fully the format and orientation of the how-to guide and the complex engagements with digital technology and politics that unfold through seemingly practical courses of action. This chapter introduces the how-to "cosmology" that informs this study's approach to practice-based research. It is written as an extended deliberation on instruments and instrumentality, interrogating how citizen-sensing practices and technologies are meant to operate and how they actually perform when put to work in the open air, thereby generating what I call *open-air instrumentalisms*.

Chapters 2, 3, and 4 narrate the fieldwork, installations, and collaborations undertaken with communities working with citizen-sensing technologies to monitor air pollution. Chapter 2, "Speculative Citizens: How to Evidence Harm," focuses on Citizen Sense's first project for sensing air pollution from the hydraulic fracturing (fracking) industry in northeastern Pennsylvania. This chapter describes attempts by residents to document effects from pollution in response to corporate and state neglect. I situate this work within citizen-sensing practices that grapple with how to generate forms of evidence while also building different infrastructures that could create more breathable worlds.

Chapter 3, "Data Citizens: How to Reinvent Rights," details residents' and workers' use of air-pollution sensing technologies in the context of rapid urban development in South East London. Citizen data become enrolled in ongoing projects that attempt to reshape and preserve the urban realm by articulating the right to breathe. Here, the right-to becomes aligned with the how-to through the collection and mobilization of citizen data. However, as this chapter suggests, data could even supplant a struggle for rights when citizens feel that rights do not provide an actionable or equitable basis for addressing environmental pollution.

Chapter 4, "Multiple Citizens: How to Cultivate Relations," documents the construction of air-pollution gardens in the financial center of London. This chapter works through the conjugations of sensing subjects that occur across humans and more-than-humans in the development of gardens that include sensors and vegetation responsive to air pollution. Sensing organisms such as vegetation can offer a way to observe and mitigate problems of air pollution. Chapter 4 describes efforts to sense and rework air pollution by transforming urban infrastructures and incorporating multiple other entities into the project of making breathable worlds.

I close *Citizens of Worlds* by considering the citizenships worked through in this book. Atmospheric, instrumental, speculative, data-oriented, multiple, and many other citizens surface here. The conclusion, "Sensing Citizens: How to Collectivize Experience," reconsiders how environmental collectives form and are engaged in ongoing if differing struggles to make breathable worlds. Before and after the four chapters, citizen-sensing toolkits present different sensor configurations that the Citizen Sense research project has studied, tested, and installed. The adjacent chapters document and analyze these toolkits as attempts to develop practices of computing otherwise.

Citizens of Worlds is a proposal to move into the open air when studying how citizens and worlds, technologies and practices, materialize through concrete environmental struggles. A turn toward the open air involves a turn toward concrete effects and practices—it is an orientation and an undertaking that might involve combat breathing as well as many other struggles for reworking the pulsations of daily life. Air is more than a volumetric container, element, or essential unit of analysis. Instead, it is a differential process, material, exchange, and condition for being and becoming citizens of worlds. Such an approach considers how political subjects form with and through exchanges that sense and build more breathable worlds. The atmospheric citizen discussed here is an initial figure that signals how citizens and citizenships form through atmospheres and injustice, air pollution and air-quality monitoring. Many more citizens and worlds materialize in the chapters that follow. This text is written toward pluralistic practices and modes of citizenship—atmospheric and otherwise—that materialize through struggles to sense and act on air pollution as a way to realize more breathable worlds.

CITIZEN SENSE TOOLKIT

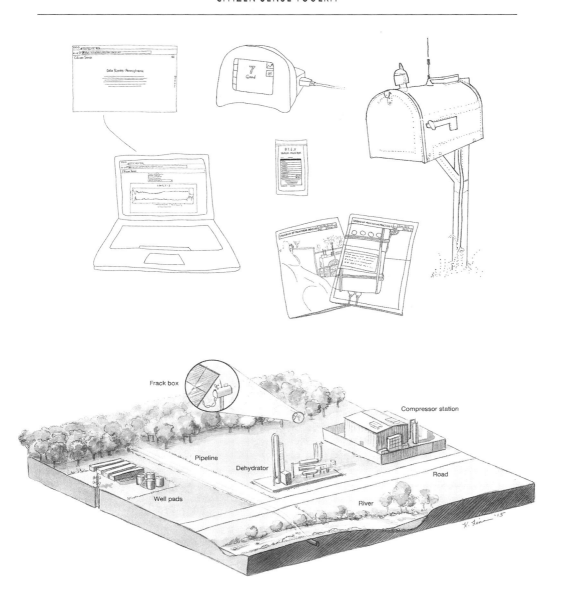

Citizen Sense Kit developed for monitoring air quality in relation to fracking infrastructure, northeastern Pennsylvania. Illustration above by Sarah Garcin, illustration below by Kelly Finan; courtesy of Citizen Sense. This toolkit can be found in a more extensive form online at https://manifold.umn.edu/projects/citizens-of-worlds/resource-collection/citizens-of-worlds-toolkits/resource/citizensense-kit.

Chapter 1

INSTRUMENTAL CITIZENS

How to Retool Action

The world of sensors is one of amplified connections. Sensors are meant to join up and speed up, while also facilitating and enabling. Whether adjusting lighting levels or advancing political engagement, a quickening of activity is expected to unfold through sensors. Sensors are embedded in urban infrastructure and surveillance systems, and they are packaged as makerly kits and citizen-sensing projects. But to get here, you need to follow the instructions. Working with sensors typically involves the uptake of the how-to guide. Citizen sensing requires settling into the instructional mode and the imperative mood. A sensing citizen is a handy subject, an action-oriented and technically equipped actor able to tinker toward new configurations. Behold, the instrumental citizen!

Sensors are a key component of citizen sensing. They enable the measurement and detection of environmental variables. Sensors, and the growing use of these devices for citizen-sensing projects, are in some ways part of a wider movement toward the how-to and the do-it-yourself. Handbooks and user's guides cover topics both technical and philosophical. From YouTube videos providing instruction on how to troubleshoot the use of microcontrollers, to handbooks for using and "abusing" the Internet of Things (IoT),[1] to how-to forums and Instructables, the genres and formats of the digital continue to expand and develop into well-used vehicles for technical instruction. How-to guides and instructions are integral to computation. Code is an unfurling of instructions. Algorithms set in place procedures for computational processes. The instructional has intensified its residency in machines.

As the how-to proliferates and instructions unfold through every aspect of the computational, this chapter considers why the how-to has become one of the prevailing genres of the digital. Alongside wondering about the how-to format, this text examines how the instructional approach to sensors contributes to particular

ways of engaging with these devices. "How to do things with sensors" is as much
a question as a set of instructions that considers how things are made doable
with and through sensors, and by specific modes of instrumental citizens. A
study of sensors could unfold through any number of modes of inquiry, whether
an ethnography of use or a technical exposition. Because sensors are meant to
organize worlds of action that include building, coding, and installing these technologies, while also analyzing data that they produce, this chapter-as-guide describes practice-based research that attempts to inhabit the practical cosmologies
of sensors. This research informs the chapters that follow, which describe how
the Citizen Sense research group set up sensors in the open air. As a guide, this
chapter further considers how worlds are made sense-able, breathe-able, and
action-able through the instructional mode of citizen-sensing projects.

Although the how-to guide is now prevalent within digital spaces, it is cer-
tainly not unique to the digital realm. Handbooks advise how to live forever,
how to live on Mars, how to meet aliens, how to conquer the internet, how to
make a million, how to build a rocket, how to clone a sheep, how to split the
atom, and how to save the planet. And this list is just a cursory scan. How-to
guides run into the tens of thousands. The how-to guide not only outlines procedures for attaining the grandiose or epic but also informs on the banal and
the necessary. There are how-to guides for foraging for food after a disaster and
learning handicrafts. How-to guides also provide instructions for organizing
political campaigns, undertaking direct action, and following step-by-step programs toward greater democratization.[2]

One subgenre of the how-to guide that is of particular interest for this study
is what might be considered a more radical or countervailing approach to instructions. These how-to guides span from instructions for surviving in a world
with which one is at odds, to working and reworking digital technology toward
engagements beyond the usual privileged actors, and making more livable worlds.[3]
These how-to guides are far from simplistic in assessing the problem at hand
or remedying approaches. Instead, they often undertake a necessary and power-
ful diagnosis of inequality, injustice, and distress and how to combat or circumvent these conditions. In this way, the how-to guide is not merely instructional
for assembling gadgets; it is also a life guide, suggesting how best to carry on.
Still other how-to guides are written as counterinstructionals. Joanna Russ's
How to Suppress Women's Writing uses the how-to guide to demonstrate the multiple strategies by which women's writing is dismissed, derided, appropriated,
and erased. Her point is not to facilitate or encourage these practices but rather
to expose the habitual, recurring, pernicious, if often implicit ways in which some
voices are excluded to amplify the writings of the privileged. How-to guides

express political commitments. They are not merely a universal set of skills for anyone to follow, even for a seemingly accessible technology such as citizen-sensing devices. They can also serve as normative devices, reproducing unjust political structures and relations. Or they can provide resources for breaking with and addressing inequities.

How-to guides are often organized as or accompanied by toolkits that serve as the essential components for configuring and materializing instructions. Toolkits are the thing to be constructed, gathered, and mobilized; they are the instrument to achieve particular effects. Astonishing capabilities are often channeled through toolkits, from suitcases and bug-out bags that can save you in an emergency[4] to ad hoc solar power setups and moonshine installations.[5] At least since the 1960s, how-to guides have been bound up with a seemingly counter-cultural ethos, where access to tools might rework living environments.[6] Some of these kits and instructions yield the unprogrammatic, the unexpected, the incommensurate, and the incomputable.[7] Other DIY projects present "tactical" strategies for intervening within the current operating systems of technoscience to realize more socially just or equitable relations.[8]

Within this instructional realm, it seems the manifesto has given way to the toolkit. The manifesto form of often dogmatic proclamations has yielded to a more open-ended organization of practices. Radical toolkits, in particular, gather instruments and resources that serve as practical philosophies and modes of organizing in difficult circumstances. There are speculative "feel-tank kits" to retool political action and imaginaries; "killjoy survival kits" for living a feminist life and surviving ongoing adversity; and radical pedagogy kits for sharing skills, lore, and community-organizing techniques.[9] At the same time and in a diverging way, policy recommendations now often circulate in the form of toolkits, a format that governments and NGOs use to advocate for the accessibility of the policy-making process.[10] This approach runs the risk of shoring up power structures while presenting governance as a more transparent and DIY process. In this sense, toolkits can be shape-shifting genres that variously provide guidance for rerouting, or reinforcing, sociopolitical practices.

While the focus here is on the how-to in relation to digital technologies, especially environmental sensors and citizen-sensing projects, an extensive assortment of instructionals and toolkits informs this study. A history of how-to and instructional guides would be an interesting project, although this is not within the scope of this research. Instead, I examine the instructions and instructional approach that inform and develop through the configuration and use of citizen-sensing technologies, which informed the practice-based methods of the Citizen Sense research project. In working through the instructional approach, I consider in what ways how-to guides not only enable technical engagements with citizen-sensing devices but also provide distinct material formats and political practices for addressing environmental problems. Toolkits are guides to action. They offer ways not just to make sensors but also to construct breathable worlds.

The procedural approach could promise an outcome of sorts. Yet what happens when instructions are imperfectly followed or ignored? And what occurs if the promised effects do not unfold as expected? The how-to guide could be a helpful set of instructions and a potentially overly programmatic mode of engagement. The how-to quickly moves from the how-to-do-it to the do-it-this-way-or-else. The how-to could become a means of cybernetic command and control: follow these steps to a certain outcome, not through an overarching program of control but the pursuit of edifying instructions. Here technology could close in on itself, and instrumentality could lead to the bad sort of functionality against which philosophers engaged with technology have warned. Technologies in this register are seen to fulfill mere functions. As Gilbert Simondon has suggested, this master–slave condition can overlook or suppress the relations that unfold across environments, subjects, and zones of energetic and cultural transfer.[11]

The how-to could, then, be expressive of an *imperative mood*. Procedure and instructions are bundled up with a pedagogy infused with soft commands and imperative verbs. The digital propagates a cascade of directives. The how-to maker-verse assembles as a hub of instructions. Image tiles organize into task sessions; search bars open up into multidimensional cosmologies of the how-to; the ten-minute step-by-step guide promises incremental if expedient accomplishment; and the soothing virtual confab of video narrators begins with the ever-familiar entry point: "Hi, guys, I'm here to show you how to . . ." The instrumental citizen who enters the how-to process is not just learning how to make and use technology but is also entering into modes of procedure, instruction, implementation, and instrumentality that guide digital participation.

This chapter engages with the how-to and the imperative mood as critical aspects of how citizen-sensing technologies and practices occur. On the one hand, it charts a how-to investigation of citizen-sensing practices and procedures. It recounts experiences of initially testing sensors and deploying these in the open air, while providing a guide to carrying out practice-based research into citizen sensing. On the other hand, it reflects on the how-to as a genre and approach meant to realize more proficient digital and instrumental citizens. Yet this seemingly straightforward practice of monitoring environments with sensors to activate political change rarely—if ever—unfolds in such a straightforward way. Nor does citizenship magically emanate from the use of these devices. The uptake of citizen-sensing devices, then, triggers a critical set of questions aligned with those raised in the introduction to this book.

Which citizenship practices do these digital technologies activate, legitimate, reproduce, or transform, and who can operate as an instrumental citizen? Suppose the how-to aspect of these technologies is meant to guide not just the construction of sensors but also the formation of political subjects. How then do sensors influence environmental actions and relations? And if the use of citizen-sensing devices does not lead to specific outcomes, how do these practices instead generate open-air instrumentalisms that contribute to struggles for breathable worlds? These questions inquire into whether citizen sensing achieves its stated results and realizes its hoped-for effects, or whether different engagements unfold along the way that rework the operations and relations of these instruments. At the same time, these questions consider who registers as a political subject and how or whether sensing technologies reproduce the inequalities that often characterize citizenship.

Rather than addressing these questions by describing the phenomenon of the how-to through a distanced commentary, I work through instructions and approaches to building toolkits, monitoring environments, analyzing data, communicating evidence, and attempting to realize political change through the Citizen Sense project. I consider how protocols and toolkits, practical manifestos and political programs, inform and materialize as citizen-sensing projects and practices. By attending to the how-to as a particular orientation to technology, I suggest that engaging with toolkits and guides is also a way of working through, reworking, and transforming the possibilities of technical, political, and environmental practice. In this way, I examine how an engagement with toolkits can become a way to *retool* approaches to instruments, instrumentality, and digital worlds in the making.

This investigation into how to do things with sensors is a modest proposal for practice-based research methods. It unfolds by recounting experiences of working with citizen-sensing technologies while collaborating with a wide range of participants. By working through engagements with sensor technologies, communities, environmental pollution, and political processes, this research suggests how to study and activate citizen-sensing actions. Practice-based research can generate techniques to query the promised effects of sensors and test forms of political engagement. Yet practice-based engagements with citizen sensing demonstrate how technologies organize action and create effects in ways that are never as straightforward as initially promised. These are practices in the making, where technologies, citizens, and political relations materialize within distinct milieus.[12]

The practices involved in undertaking citizen-sensing projects require not just putting together assorted electronics but also attending to complex configurations of technology, politics, environments, and modes of citizenship. This examination of the how-to and the toolkit does not idolize making and craft but instead offers a theory of practice and action oriented toward the possibilities for building more breathable worlds. This text attends to what is often left out of how-to guides. By engaging with what might be on the margins of technical interest, I hope to rework how to encounter citizen-sensing technologies, less as instruments able to implement certain ends and more as openings into rethinking socio-environmental potential and technopolitical relations. Can instruments generate instrumentality, not of the positivist sort, but more of the pragmatist type, where, as John Dewey has suggested, instrumentality necessarily requires an experimental and contingent set of engagements?[13] While such an orientation might seem to point toward instrumental relations in the sense of easy outcomes, it instead gives rise to expanded engagements with action and effect. Instruments as ideas require what William James has called the "open air" to unfold as lived experiences and processes.[14] Indeed, as Cornel West has pointed out, pragmatism concerns the contingencies of subjects, collectives, and worlds as well as theories and knowledge, which in their more pliable formation can act upon and respond to social crises and democratic struggles.[15]

By reading pragmatism sideways through engagements with feminist technoscience, environmental justice, Black studies, and Indigenous theory, and by working with instruments in practice, I suggest that it might be possible to re-engage with instrumentalism beyond its usual extractive and expedient registers to consider expanded relations of effect and effectiveness. I situate this engagement within the open air of inquiry to express the sociopolitical constitution of instruments as much as to relay how devices are situated in and make worlds. The how-to is a proposition for open technology, which can be a way to engage with machines beyond fixed outcomes.[16] How-to here becomes an invitation to make, organize, orchestrate, conjure, and sustain people, technology, and worlds toward openings rather than prescribed ends.[17] While ends might inform the starting points for particular technopolitical practices, they inevitably change along the way. An instrumental proposition becomes a site of transformation. This chapter-as-how-to-guide proposes how to retool the how-to approach, not to proceed toward certain results, but rather to work for open engagements. As noted in the Introduction to this book, I call this approach *open-air instrumentalisms*.

Retooling is a practice in open-air instrumentalism that creates and tests actions for working toward more breathable worlds.

HOW TO CONSTRUCT TOOLKITS

Many digital technologies occupy a curiously contrary position. They seem to offer engagement and empowerment for individual users. Yet they are often highly controlled and monopolized technologies that bind users into particular practices and relations. Such an observation has been made through numerous analyses of an array of digital gadgets, data, and platforms.[18] Digital technologies organize deceptive relations to a seemingly empowered if minutely surveilled user. Social media enables views to be broadcast while profiling users. Apps and wearables track sleep and fitness while promising greater well-being and productivity. Digital technologies manage, inform, and otherwise mediate everyday activities, and these functions are overseen by a limited number of organizations with often questionable agendas and discriminatory practices.[19]

To unravel the typically rigid contours of digital devices, many DIY technologists have begun to assemble microcontrollers, sensors, and code to develop a more informed engagement with these machines. Here the toolkit becomes a way to fashion a more deliberate encounter with digital devices. By making electronics from the ground up, one is meant to decode the decisions made in setting up technical configurations one way and not another. But even the fashioning of a toolkit already contains a set of built-in assumptions and orientations toward technopolitical action. Tearing down electronics and remaking them into toolkits for assembly is a process I examine more closely here. Multiple toolkits address how to develop observant digital practices, including how-to guides for erasing your internet profiles, how to become anonymous online, how-to instructions for undergoing a data detox, and how-to guides for building a DIY bulk surveillance system.[20] In this sense, toolkits not only provide instructions and materials but also indicate how to live with, through, and against these technologies. Toolkits provide instructions not just for assembly and use but also for attending to the social and political ramifications of digital devices.

Citizen-sensing technologies present a similarly complex set of instructions, practices, relations, and politics. Many of these technologies monitor environmental variables, such as air or noise pollution. Air-quality monitoring toolkits can be found in multiple forms, which do not even necessarily include digital devices. Indeed, there are long-standing practices of working with diffusion tubes to monitor nitrogen dioxide. Small plastic tubes can be affixed to lampposts and street

Figure 1.1. Prototyping a first-generation AirCasting air-quality monitor from Habitat Map. Photograph by Citizen Sense.

Figure 1.2. Testing a plug-and-play Airbeam air-quality monitor from Habitat Map. Photograph by Citizen Sense.

signs across an urban area and then sent to a laboratory for analysis. Such low-cost analog devices provide a monthly average of nitrogen dioxide levels and indicate pollution levels in an area. These monitoring practices often also include instructions for how to undertake a campaign for improving air quality, together with technical instructions for installing diffusion tubes and analyzing data. The expanded toolkits that mobilize along with these devices can be oriented to community organizing and local activism, urban design and traffic interventions, collective mappings and town hall meetings, and problems of digital functionality. These aspects of toolkits are no less important, yet they tend to recede from view when the focus is on learning the technical aspects of digital monitoring technologies.

The rise of citizen-sensing practices and technologies could in one way seem to activate instrumental—or, in other words, potentially reductive and functional—approaches to citizenship and political engagement.[21] Citizen sensing is a collection of technologies and practices for monitoring the air that are often bundled into toolkit form. These digital technologies are used to monitor and measure environmental problems and to generate data that could be actionable for policy and regulation. Yet in another way, these instruments in the form of low-cost environmental sensors could rework what might be seen as instrumentalist approaches to politics. They could develop different vocabularies and trajectories of effect and effectiveness that challenge the apparently linear logic of these instruments. By following the seeming instrumental course of the how-to, citizens could find they are instead involved in more knotty and wayward struggles for more breathable worlds.

Getting Started: An Incomplete List of Sensor Kits

Let's look more closely at a few citizen-sensing technologies to consider how these practices present alternative strategies for documenting and acting on environmental problems. While only a few of the many sensors available for environmental monitoring are addressed in this how-to investigation, there is an extensive array of citizen-sensing projects that have been variously reviewed, assessed, tested, and assembled throughout this research.

The focus here is on monitoring air quality. Yet there are many more sensors for monitoring water, noise, vibration, temperature, humidity, wind, heat, energy, radiation, soil, and vegetation. Water quality can be monitored through conductivity, temperature, and total dissolved solid sensors like the CaTTfish, and water levels can be monitored through ultrasonic sensors like the Flood Monitor.[22] There are Pocket Geiger sensors for measuring radiation,[23] and DIY seismic sensors for measuring earthquake activity.[24] Moisture sensors monitor the health and presence of vegetation,[25] and temperature and humidity sensors monitor beehives and ensure the health of honeybees.[26] One of the earliest and longest-standing uses of DIY sensors has been monitoring air quality. Numerous sensors are now in circulation for monitoring air, from the DIY to those sold as finished products.[27] A partial list of operational, obsolete, and speculative citizen-sensing devices and toolkits for monitoring air quality includes the following:

Aclima	hackAir
Airbeam	IGERESS
Airbox	InfluencAir
AirCasting	IQAir
AirKit	iSPEX
Air Quality Egg	LaserEgg 2
AirSensa	LifeBasis
Air Sensor Toolbox	LondonAir App
AirVeda	Luftdaten
AirVisual	MicroPEM
Alphasense Sensors	Netatmo
Area Immediate Reading (AIR)	NOKLEAD
Array of Things	PANDA
Atmotube	Plantower
Awair	Plume Air Report
Breathe Cam	PuffTrones
Brizi	PurpleAir
Cair Smart Air Quality Sensor	PUWP (Portable University of
Citizen Sense Toolkit	Washington Particle monitor)
CityAir App	Safecast
Clarity	SensorBox
Clean Space Tag	Sensors in the Sky
Common Sense	Shinyei Particle Sensor
DR1000 Flying Laboratory	Sidepak Personal Aerosol Monitor
Dustbox	Smart Citizen Kit
DustDuino	Smoke Sense App
Dylos	Soofa Benches
EarthSense Zephyr	Speck
Float_Beijing	Tree Wi-Fi
Flow, Plume Labs	Tzoa (Enviro-tracker)
Foobot	WeatherCube
Grove Air Quality Sensor	Wynd

These are DIY sensors and off-the-shelf kits, wearables and desktop devices, as well as a few apps that double as sensors through the use of smartphones for monitoring or navigating environments. The sensor names variously suggest democratic initiatives, technical enterprises, environmental revitalization, research projects, and manga characters. The list of names further designates projects and products, communities and practices, locations and platforms. In other words, these toolkits do not de facto include certain components and exclude others. They are in varying states of composition and decomposition, salability and availability, with different monitoring capacities.

Some of the companies or makers of the sensors mentioned above make avowedly apolitical statements, which indicate that the data the sensors collect are not intended to support political projects, nor are the data open to the collectors of air-pollution data.[28] Other projects, such as the Dustbox developed through the Citizen Sense research group, deliberately seek to investigate how or whether new types of political and environmental engagement can materialize with these sensors.[29] Many of these projects focus on the technical device that is meant to organize, attract, and mobilize citizen participation and data collection. The problem of air quality is in part organized through the devices and practices that environmental sensors make possible. Monitoring air pollutants, collecting data, and communicating evidence about elevated pollutant levels could be seen to be instrumental approaches to air quality, which these instruments facilitate. The instruments and instrumentality of air quality unfold in relation to this broader proliferation of sensors and sensor citizenship. Yet these are open-air instrumentalisms that do not follow a unilateral trajectory.

In this incomplete list of citizen-sensing projects and technologies, you might notice that some devices are ready-made and others require assembly. A number of devices are "locked down" as consumer products while others require ongoing tinkering and maintenance. Several of the sensors included in the preceding list are beta-stage and prototype technologies that have particular idiosyncrasies and require adjusted setup, troubleshooting, and puzzling over how to work with the data they collect and present. Throughout this research, sensors available in makerly form have increasingly crept toward a more settled, product-like state, and in the context of the IoT, several plug-and-play sensors are now available. Yet, if you have worked through building and setting up air-quality sensors, you are inevitably left to wonder about how or whether such off-the-shelf devices are calibrated, how to access and analyze data, and whether data can be used to support claims about environmental pollution.

This overview of air-quality devices and practices is also incomplete because sensors, and especially air-quality sensors, continue to multiply and expire within the usual fleeting time spans of electronic devices. As soon as sensor technologies and projects are identified, new devices emerge and others lapse into obsolescence. Some of the air-quality sensors in this list are new, some are prototypes no longer in use, and some are dead devices that would require considerable effort to reboot and plug back in to operable systems. As with many tech projects, once devices sediment into stable forms, they as readily disappear or cease functioning, with websites flickering into oblivion, firmware updates colliding with hardware configurations, and peripherals no longer communicating across ports.[30] But this observation also jumps ahead, because it points to a few things to keep in mind when starting to use citizen-sensing toolkits. First, it is worthwhile to discuss briefly how sensor kits are configured as cosmologies of sorts.

Flat-Pack Cosmologies

The modular instructions and diagrams for assembling toolkits demonstrate a distinct approach to problems, where relevant components are gathered together, documented, and assembled into an entity that will address the problem at hand. Think of the flat pack that consists of an itemized inventory of parts, including atomized images of assembly, with connecting actions signaled through arrows

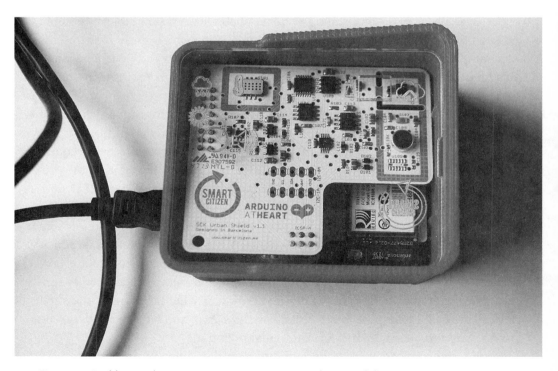

Figure 1.3. Building and setting up a Smart Citizen Kit. Photograph by Citizen Sense.

Figure 1.4. Unboxing and setting up the Air Quality Egg. Photograph by Citizen Sense.

segueing across framed sequences toward a clear outcome. Similar to many modular products that can now be purchased and assembled with apparent ease, a certain flat-pack relationality is operationalized through sensor toolkits. All the items needed to complete the project need only be joined together by following instructions. What begins to unfold in this approach to toolkits and instructions, actions and outcomes are flat-pack cosmologies, where the speculative pluriverse of environmental problems assembles into neat diagrams of constructable relations held together through air-quality sensors.

Cosmology is a term used by Alfred North Whitehead to describe a metaphysical system that represents a universe of relations undergoing processes of transformation.[31] The term also has a longer history of use within Indigenous theory and practice to refer to experiences, connections, and activities held in common.[32] And Alexis Gumbs's breath-based cosmology, discussed in the Introduction, suggests how to cultivate connections for realizing political possibility. While *flat-pack cosmologies* is inevitably a diverging use of the term and concept, it indicates how the toolkit as a distributed and connected system forms and works, including how entities develop, how relations join up, how societies materialize, and how these varying components unfold and are sustained because of the values attributed to technology, especially in modular form.[33] If there is no one-world world, by extension there is no one-cosmology cosmology. Cosmological systems of relations form through experiencing subjects, whether sensors or humans, which also have political effects.[34]

These cosmologies provide reference points for engaging with the relations that toolkits call into being. Toolkits offer a particular way of concretizing maker-subjects, items for construction, modes of assembly, models of action, modes of becoming, desired outcomes, and strategies for holding things together or throwing everything away. Toolkits are ways of organizing problems and relations for action. Even when the point is to crack open or even tear down the black box of sensor technology,[35] the component parts indicate new modes of assembly. This tearing down and putting together, constituting and reconstituting of entities, is indeed a salient characteristic of toolkits, which are notable for their modularity and flat-pack-ness but need to be sufficiently open to be adaptable to new circumstances and uses.[36]

Toolkits, especially air-quality monitoring toolkits, are at once procedural and contingent arrangements. Indeed, the procedural method becomes quickly troubled in the flurry of calculating how to work with toolkits in the open air of environmental monitoring. The next section details these points of procedure as well as how they can go awry. Also of note with the how-to guide is that the you/your of the instructional refers to a maker who is brought into a technical relation. As I follow this instructional, I similarly traverse from first to second person to inhabit the you/your of the instructional and the imperative mood, and to work through the practice-based and participatory aspects of setting up DIY sensor technologies. Let's begin.

Ten Points for How to Construct Toolkits

When starting off with a citizen-sensing project, one of the first things you might wonder about is how to construct a monitoring toolkit. Yet a toolkit is always more than just a collection of sensors and assorted digital gear. Some toolkits more extensively outline techniques for learning protocols, organizing collection efforts, analyzing data, and influencing policy than the finer details of technical kit configurations.[37] Here are a few notes that outline some of the key considerations when beginning your project. Afterward we will look at a few sample projects that will give you a more detailed sense of how these points could be implemented.

1. What is a toolkit?

The first thing to keep in mind when doing things with sensors is that these typically makeshift instruments will give rise to questions about the purpose, the composition, and the coherency of the technology under investigation—but never in such a philosophical way. Instead, the question will arise in the middle of attempting to get sensors to work. The refusal of electronics to function as a key part of a sensor toolkit will reveal the limited effect and scope of these devices. The eventual functioning yet occasionally inexplicable output of sensors will make one wonder at the apparent achievement of obtaining a connection. Sensors have not necessarily, as Simondon would suggest, become sufficiently integrated so as to seem "natural" within their own self-generated milieu.[38] Instead, they are often troublesome contraptions that consume your time and energy as you attempt to find an operative pathway to citizen engagement. Sensors, you might discover, are just one particular entry point for engaging with air pollution, which is also interconnected with environmental public health, development, community organizing, and environmental justice.

2. Which sensor should you use to monitor the air?

This is a question I am often asked by people interested in beginning to monitor with sensors. The answer is, it depends. This is the second note on how to do things with sensors. The sensor you choose to use depends on whether your interest is to tinker with electronics, to plug in a device without having to modify it, to focus on collecting "accurate" environmental data, to map and share data with a wider monitoring community, or to focus on a particular air pollutant of concern. These are not always mutually exclusive objectives, but often the focus will be placed on one priority area more than another.

3. What parts will you need?

The third note about how to do things with sensors is that most guides will begin with a seemingly comprehensive list of parts. Photographs of parts show neatly arranged and brightly colored LEGO-like electronics that beckon for a makerly connector to join them up. The parts will include jumper cables and wire ties, Velcro and tape, breadboards and LEDs, microcontrollers and gas sensors, potentiometers and buzzers, 9-volt and lithium batteries, and resistors of various sorts. This list of parts is assembled in different ways, but the basic components include electricity and computation, held together through digital infrastructures. This is less a foundational universe of the four elements and more an operative pluriverse of the many effectivities. As noted, the how-to pluriverse assembles through flat-pack cosmologies. These cosmologies seemingly include all the parts and instructions you will ever need to realize your objectives. However, just as your cosmology of electronics begins to assemble, you will discover that a part is missing or that the comprehensive list of parts defines sensing in one way, such as how to pass voltage along a wire, and not another, such as how to convert voltage into a semi-accurate measurement of pollution levels. The cosmologies of the flat pack are always in process, splintering into multiple cosmologies of what the how-to kit could enable or open up.

4. Where should you begin?

The fourth note when working with sensors is that you are most likely better off diving in and tinkering with a bit of kit before you assiduously read the instructions or absorb extensive advice. The intricacies of pins and holes, cables and ties, are best encountered through physical proximity rather than secondhand reports. Turn to forums and videos once your brain is on the bake, the electronics refuse to talk, and you are sufficiently prepared for the curious if unique hybrid of geekery and spleen that often pours forth from makerly FAQs. Your virtual interlocutors will frequently declare, "No, I will save the planet first, and it will only be through my bespoke circuit diagram!" The how-to should, for this reason, be approached with caution in the face of such zealotry.

5. How do you make a working sensor?

The fifth note on how to do things with sensors is that the device you are working with will likely need upgrades and updates before you have even begun. The instructional guide you follow will recommend software or hardware that is unavailable or out of date. The microcontroller hardware will have been updated to a newer version. You can start with an Arduino microcontroller, only to discover that the software libraries to be loaded on your Arduino no longer function with your microcontroller version. The entire configuration of the sensing kit will have shifted so that a new iteration needs to be developed through the very making and following of instructions. To do things with sensors, you need to trudge through states of nonconnection and electrical blank spots. Lights will refuse to flash, data will decline to post, and URLs will flash "404" where platforms should appear. The online forum and the FAQ section will become your most helpful resources in these early stages. There you will find the near-time updates and fixes that will come to your aid as you bodge your way toward a working sensor.

6. Are we there yet?

The sixth note is that a sensor toolkit will never be complete. It is a roving arrangement of stuff that will need to be topped up, updated, supplemented, extended, and hacked together. A tidy toolbox will soon become the site of a mass spill event. A clear desk will conceal an essential cable. Online warehouses will become ever-expanding otherworldly depots, where making and remaking require just one more trip to the webby aisles of Cool Components. Where does the necessary kit for undertaking a citizen-sensing project begin and end? This question could forever remain unanswered.

7. What should you do with the data?

The seventh thing to consider is that once you have built your sensor, you will need to post your data and map your monitoring locations. This process can occur on a platform that you develop along with a sensor device or on an externally developed platform that may or may not last the year. Once platforms expire, your device will likely require an entirely new configuration to pipe data to another platform. Some platforms require inputting latitude and longitude to mark fixed locations. Other platforms track the routes and itineraries of sensors used in more mobile ways. Platforms can also include the outputs of sensor data, which are presented in a wide variety of formats, from raw voltage counts to units converted to regulatory standards of parts per million (PPM) or parts per billion (PPB) or micrograms per cubic meter ($\mu g/m^3$). Pollutants monitored could include gases and particles, from nitrogen dioxide to particulate matter. These different ways of presenting and engaging with data raise multiple other how-to questions about how to analyze data sets, how to generate evidence, how to communicate findings, and how to influence policy—topics I address in more detail in the chapters that follow. A successfully connected sensor, as it turns out, will be the least of your problems once you move to the domains of platforms and data.

8. How should sensors be used in the field?

Now that you're well on your way to making your first citizen-sensing toolkit, here's the eighth thing to remember: if your plan is to monitor environments, you will eventually have to move from the desktop and workshop to the open air, where sensors will be used in situ, over time, and in a range of conditions. You might have bashed together a passable device consisting of a metal oxide sensor, breadboard, jumper cables, and resistors, all activated by a lithium battery. You might even have mapped or located your device on an online platform, thereby giving your project an apparently global reach. However, there are yet more instructions that could be written for how to undertake pollution sensing, which necessarily expand once sensors are taken out into the open air. This guide attempts to account for the contingencies, experiments, and openings that occur through such open-air instrumentalisms.

9. How do you register pollution?

Assuming that the purpose of putting together sensors is to monitor environments, then a toolkit assembled for addressing air pollution will also ideally need to address practices of environmental observation and engagement. Note 9, which could even precede note 1, is that although it can be quite easy to get bogged down in making sensor devices work, the milieus of these technologies necessarily comprise events such as how pollution registers or might register—including in existing monitoring networks, asthmatic bodies, or regulatory violations—and how communities become involved in attempting to address the problem of air quality to make more breathable worlds. The scope of your toolkit might need to be redrawn so that the environmental, social, and political aspects of monitoring are as much an area of study as attempting to create a blinking LED.

10+. How do you create a community-monitoring project?

Note 10, sprawling to an indefinite number of notes, is that you will encounter endless considerations for how to make sense of your attempts to monitor air pollution once in the open air. This list of further things to consider includes: Which air pollutants are you monitoring? Where are the likely sources of emissions? Which monitoring protocols and methods will you follow? Have you calibrated your sensor? How often and for how long will you collect data? Will your monitor be stationary or mobile? How many locations are in your monitoring network? How will you compare data across monitoring locations? How will you compare your data to other regulatory or reference monitors? In which measurement units will you present your data? How will you analyze your data? With whom will you share your data? What do you hope to change, improve, or challenge about monitored environments? With which organizations or regulatory bodies might you collaborate to act on findings from community monitoring? Think of these points as a guide that can be read alongside examples of citizen-sensing installations and in relation to your own monitoring project.

Although this section is first arranged as a ten-point plan in keeping with the exigencies of the how-to genre, it quickly unfolds into an open-ended set of considerations when attempting to monitor pollution. As this preliminary list for assembling citizen-sensing toolkits demonstrates, any actual monitoring project will encounter multiple situations that deviate from the instructions. These tips could be presented as a set of instructions or as a checklist for how to go about monitoring, but they are also far from definitive in terms of addressing the particular conditions that could arise when undertaking environmental monitoring.

In the process of following these instructions, you might have learned that the assembly of sensors is far from straightforward, the composition of a toolkit is neither fixed nor complete, the posting of sensor data can come in many forms, the analysis of data is an area of ongoing development, the protocols and methods for monitoring are often still in process, the "citizens" who would monitor are often differentially able to make their voices heard, and the environments to be monitored will make specific demands upon how data are collected, presented, and turned into evidence. When following any instructional guide for sensors, you will find that open-air instrumentalisms abound.

Yet this is not to say that it is impossible to operationalize environmental sensors to detect pollution and gather data. Multiple low-cost and DIY sensor projects are now in place that continually collect data and document environmental processes. However, these practices are provisional and full of necessary workarounds.[39] Inevitably, more than a few how-to guides for using sensors will present this as a simple and matter-of-fact process.[40] I suggest, conversely, that by attending to the deviations from the straightforward approach, you could find that many more engagements with sensors, environments, and politics emerge that remake the operations of instruments and instrumentality. Open-air instrumentalisms in this way are to be valued, because they are the process through which technopolitical experiments and more just environmental collectives could coalesce. I will have more to say about this point in the sections that follow.

Now, let's turn to look at some detailed examples and experiences of assembling sensors for testing and eventual use in the field. A few things to ask along the way are: How does a toolkit expand or shape-shift along with differing uses? What other considerations come to the fore when attempting to monitor air pollution near an industry site or in a congested city? And in what ways does the how-to expand from technical delineations to indicate that these practices have been political all along?

HOW TO CONNECT SENSORS

It could seem straightforward enough to buy a sensor, plug it in, and begin monitoring. But the situation is rarely as simple as that. The process of working with sensors includes struggles with upgrades, deciphering of data outputs on platforms, trials to test sensors in situ, and uneven comparisons with regulatory monitors. Through many practical tests of off-the-shelf citizen-sensing technologies, it also became apparent that users of these devices also contributed to their ongoing development by getting them to function and by contributing to (online) communities providing mutual instructions and tips for troubleshooting.

I should preface this more detailed account of how to make things with sensors by confessing that I am not, of all things, an inveterate maker. I have attempted to cobble together windmills and model cities, jelly rolls and button-down shirts, with each object bearing the sad signs of absent hand–eye coordination. While others might have tinkered together amateur radio sets or carved out three-legged stools, for me the world of "making" has been—and remains—an ongoing challenge to align bodies, forms, and functions. This how-to guide does not unfold as the advice of a seasoned expert to an audience of eager trainees (which itself is a highly gendered way in which the digital world churns as an ongoing performance of master-y). Instead, it is an account of dogged persistence and muddling along, of the just good enough and the bang-it-together, of the electrical tape and the makeshift arrangement. Luckily, what the hand does not or cannot make in all its supposed authenticity, the computer can readily press out through a bit of code, CAD, and 3D printing, along with numerous tutorials, collaborations, and conversations. In this sense, I engage with the world of DIY sensing as one is meant to: as an amateur connected to extended communities of practice. Indeed, the DIY aspects of craft, making, and tinkering can generate different experiences of embodiment, the everyday, and collective politics.[41]

In this way, I work through the faltering processes of getting sensors up and running in a citizen-sensing research group. By "up and running," I am referring not just to making photoresistors blink and microcontrollers talk but also to the extended sociopolitical and environmental relations, from communities organizing to address pollution to participatory research practices that recast the usual contours of inquiry. But this account is still not a tale of salvation through making—digital or otherwise. Instead, it is a faltering if candid encounter with the promises of DIY. It questions the processes of making kits meant to empower while toppling prevailing power structures. By taking up tools—specifically, DIY sensors for monitoring environments—it is possible to describe the finer details of how toolkits assemble. These practices are situated ways to understand the call to "hands-on" action that is meant to remedy contemporary malaise. This approach involves looking at the work-arounds used to make toolkits operational, whether for hobbyism or environmental activism. It also describes how these technologies enable particular political engagements and ways of being and becoming citizens. Along the way, this work bypasses Heideggerian hammers to rethink and rework what counts as making along the lines of what Elizabeth Povinelli has suggested can involve a probing of differential ways of being in the world.[42]

These accounts are a small selection of the many sensors reviewed, tested, and built. They work through the how-to, document experiences with following instructions, and elaborate on the potentials and pitfalls of citizen-sensing technologies.

While these sensors were collaboratively assembled as part of a practice-based research process, it should also be said that some of these kits were assembled and tested in response to pressing public events and participatory workshops in planning. In some cases, this testing and assembly process involved flying somewhat blindly into the world of sensors. These accounts re-create the steps of testing these devices after fumbling through making and setup. In the process, I also argue that there is much to be said for using toolkits inappropriately and incorrectly. This is less a condition of embracing "error" as such and more a way of seeking out the practices that proliferate on the edges of straightforward instructions. As many feminist writers have noted about their toolkits and survival guides, it is through these processes that other purportedly illegitimate or queer ways of being in the world are forged or claimed.[43]

Making in the Imperative Mood

There are numerous guides for working with sensors, including such texts as *Getting Started with Sensors* and *Environmental Monitoring with Arduino*, among other online manuals and tutorials. These texts could seem to be a good place to start, since they set out instructions to follow, sensors to make, and ways of toggling across making and essential concepts. The process of following one of these texts, however, yields unexpected processes of making, inquiry, and engagement. If you begin your voyage with sensors in this way, you will notice certain abiding themes for assembling technologies.

See, connect, attach. Insert, double-click. Orient, insert, connect, push, fold, twist, insert. Grammarians would parse this as the imperative mood. Action verbs and commands distinctly characterize the how-to of DIY electronics. Words order action. Language exhorts. Do this and complete that. Command equals outcome. Make yourself into a model citizen, and a citizen able to make models. Along the way, there will be helpful tips and conversational sidebars to review key milestones and to assure you that the makerly relationship is more friendly than authoritarian as you progress toward your goal.

The how-to guide, then, is a genre of sorts, not only in the stylistic sense but also as a way of organizing anticipation. As Lauren Berlant has noted, "genres provide an *affective expectation* of watching something unfold, whether that thing is in life or in art."[44] What is expected in the how-to guide, and how is it meant to unfold? One could say it is a genre of problem solving. Problems are identified that can be addressed through learning and sharing skills and procedures. Yet the act of constituting problems is also a way of constituting worlds.[45] To make sensors, write code, and collect data are to undertake practices of environmental monitoring that commit to a specific way of acting on pollution. The how-to guide is meant to make acquiring and executing these practices more doable. It also creates practices that attach citizen makers and citizen sensors to distinct ways of engaging with worlds.

In the primers to citizen sensing, instructions are found in the form of how-to guides for making kits from assorted electronic peripherals. They also assemble as step-by-step directions for setting up and plugging in an off-the-shelf sensor to an online platform to view data. There are instructions for following monitoring protocols, instructions for calibration, and instructions for installing sensors in polluted locations. In the O'Reilly text *Getting Started with Sensors*, which I detail here as the first example of attempting to work with sensors, hypothetical maker–readers begin the project of assembling sensors while also "bending technology to [their] will" to control computation and environments.[46] Here the instructional promises a curious mastery of technology, which, as one quickly discovers, can be a bit misleading.

According to this basic guide, the process of getting started with sensors first involves asking "What are sensors?" The text readily provides the answer that a sensor is an electrical input device that "evaluate[s] a particular stimulus within the environment."[47] It would seem that any change or disturbance in an environment could be detected and transmitted into digital form. Sounding in a Simondonian register, the authors describe this as a process of "Transduction!" to explain how sensors control circuits and, by extension, environments.[48] Here, environmental phenomena are undergoing conversions into electrical and digital outputs. The "how" prevails over the "why"—technology performs its own logic of execution.

Yet, when following instructions in a guide and toolkit such as this one, you will find that time for reflection is often cut short, since the point is to get on and make things. After asking the maker-to-be to reflect on what sensors are or might be, the authors of this instructional abruptly rejoin, "Enough discussion— it's time to build!"[49] And so you will be off, testing batteries and breadboards and LEDs, switches and alarms. There are many sensors to build here, including infrared proximity sensors, rotation sensors, photoresistors, pressure sensors, temperature sensors, and ultrasonic sensors. It will take some time to get to the finer points of connecting up air-quality sensors, which are not even covered in this basic text. However, as you work through the configurations of these many sensors, you will find that it is one thing to generate a reading from a temperature sensor and quite another to know whether the sensor is providing a verifiable measurement. In the course of setting up temperature sensors in our work space on the tenth floor of an aging office building without air-conditioning, for instance, we found that according to our temperature sensors, indoor summer temperatures instantly leaped to 40 degrees Celsius. Was this due to the electrical wiring of our sensor circuit, was it due to the placement of our sensor near a window on the sunny side of the building, or were we really just about to perish from heat exhaustion? Sensors at this stage of assembly can give rise to extensive questions about the state of the surrounding environment.

The temperature sensor test that *Getting Started with Sensors* provides is to move the sensor in and out of the refrigerator, placing it in room temperature and then cooling it down in an appliance. But if you also want to speed up the process, you can expose the sensor to ice cubes, the text suggests. The introduction of a stimulus is often used to see whether a sensor is reading. An air-pollution sensor can be exposed to a lit match, cigarette, or vacuum cleaner to see a spike or dip in the data. The first stage of connecting a sensor, then, often involves working through these processes of setting up the sensor circuit configuration, loading a bit of copy-and-paste code to a microcontroller, and then testing whether the sensor detects the introduction of basic stimuli by generating detectable changes in the data.

Many of the kits that allow you to get started in testing sensors in this way are now available as maker kits with all the necessary parts to develop a basic plant-watering system, gas sensor, or temperature sensor. Companies such as Seeed Studio sell an array of such kits that fit within the language of other assembly-based hobbies. Parts to be assembled are included along with instructions, and the process of making is meant to generate new understandings of technology through doing. DIY practices on one level are meant to "challenge traditional hierarchies of authority and the existing status quo," as Matthew Ratto and Megan Boler suggest, by decentering the usual sites and practices of making.[50] Yet DIY can also reinforce particular ways of engaging with technology, for instance, as a project of following instructions to bend technology to one's will, or in other words, to gain technical mastery and to work on a universal if abstract problem. Mastery as a project has come under fire not just from philosophers of technology such as Simondon but also from postcolonial and decolonial thinkers like Julietta Singh, who suggests that a project of "unthinking mastery" can be a way to undo the estrangement that comes with these forms of relation.[51]

DIY practices tend to be at once open-ended and instructional. What I am calling the imperative mood in this context is inevitably related to that better-known theory of the "performative mood" developed by J. L. Austin in his study *How to Do Things with Words*.[52] The performative mood is a way of mobilizing actions, relations, and worlds through speech acts. Austin was interested in studying ways of doing things with words by considering the infelicities, misfires, and miscalculations that occur within the performative mood.[53] His theory has in turn influenced thinkers like Judith Butler, who has further investigated how social constructs like gender are performed and materialized.[54] Karen Barad draws on and reworks Butler's discussion of performativity by adding a material and posthuman dimension that shifts discursive statements to a field of multi-agential possibilities (rather than an exclusively human utterance or action). In this way, performativity is less about language in abstraction and more about what Barad calls the "conditions of mattering."[55]

Theorists of digital technologies and digital citizenship have built on these concepts of performativity to analyze how digital practices can constitute ways of performing digital citizenship.[56] Citizenship in these renderings is always a practice on the move, where political subjects and possibilities form and transform through lived engagements. While there is much more to say about performativity than space here allows, this proposal for an *imperative* mood attempts to work with and alongside this constructivist approach to words, actions, and materialities to investigate how such a mood might generate its own distinct configuration of instructions, relations, practices, technics, and milieus. Even more than drawing out the generative aspects that characterize theories of performativity, I would suggest that the misfires and miscalculations come to the fore just as readily (if differently) when engaging in the imperative mood. These misfires could also be a particular entry point into understanding how open-air instrumentalisms take hold, as swerving experiments with instruments. A resistor inserted into the same pin of a breadboard as an LED or sensor can easily end up back to front. A sensor-wire-battery configuration might connect up or become disconnected. A sensor output might be all but inexplicable. Hence every list of imperative commands comes with its inevitable section on "Troubleshooting" to help makers figure out what has gone wrong along the way. But this process plays out in much more elaborate ways than simply faulty wiring or misaligned sensors. It can also extend to dodgy code and incorrect conversions, as well as data platform errors and bungled sensor housing. Misfires and miscalculations can and often do extend to botched sensor installations, inscrutable data output, and indifferent responses to data gathered.

Many theorists have discussed how instruments generate more-than-descriptive engagements that enact worlds. In other words, instruments are world-making. They are constructive and performative of the worlds that they would detect, measure, and act upon.[57] But the imperative mood designates explicit actions along with observations that might be achieved. It constitutes the methods by which such constructions and performativity occur or falter. The imperative mood constitutes the conditions, subjects, and environments in and through which a project is meant to happen and an outcome achieved. It tends to be normative in its register of address. For example, when you load code onto a microcontroller, there is little sense that there are multiple ways of completing this task. Instead, you pursue the project in the seemingly correct way, which you are meant to master before moving to the next step. This configuration of technology, maker–citizen–subject, technical relations, and practical action is what the imperative mood designates.

This initial example of working with sensors by following a standard maker text such as an O'Reilly guidebook demonstrates a common entry point and process whereby sensor instruction occurs. Sensor ontologies quickly give way to flat-pack cosmologies, where component parts join up to create electrical arrangements of action and reaction. The progression through a guidebook and the formation of your own toolkit can be delineated as the working through of devices: from LEDs to temperature and pressure sensors. In other words, guidebooks can promise makers a form of technical mastery without considering the open air where sensors would circulate and shape-shift through attempts to monitor and act on pollution in order to build more breathable worlds. When making citizen-sensing toolkits and following instructions for assembling sensors, it is important to consider how these instructions organize a particular way of encountering environmental pollution through digital sensing, which in turn establishes technological relations and possibilities.

Simply Connect

Once you've attempted to assemble sensors from their basic component parts, you might next decide to test an off-the-shelf sensor that does not require an intricate process of assembly with breadboards and jumper cables. As noted, an increasing array of sensor objects and off-the-shelf products can be procured through Kickstarter pledges and online shopping. The Air Quality Egg, which I detail here as the second example of attempting to work with sensors, is perhaps one of the most iconic of these citizen-sensing devices. Having been prototyped through a series of hacker events from 2010 to 2012 (which I describe in *Program Earth*),[58] the Egg moved from prototype to salable product by 2013, when the Citizen Sense research project was underway. Because the Egg was an air-pollution sensor that could be purchased as a complete product, it did not require soldering or microcontroller setup. Such an off-the-shelf sensor would seem to allow for more attention to be given to environmental monitoring, data collection, and public engagement. So our research group began to investigate the capacities of this device.

We placed an order for our own Egg in the middle of July 2013. In late July, the sensor arrived at our offices in London, sent from Wicked Device, based in

Figure 1.5. Troubleshooting the Air Quality Egg, including viewing tutorials to upgrade the device middleware to make Eggs flash in different colors. Photographs by Citizen Sense.

Ithaca, New York. On its outer label, the neatly packaged device promised to make more engaged citizens of us all. It declared:

> Problem Solved. Do you ever think about the air you breathe? It affects us in ways we can see and also in ways we can't. The Air Quality Egg is a project working to make the air we breathe more "visible." Simply hang it in your home, office or outside your window to start collecting your personal air quality data. The Air Quality Egg connects you with a global community of concerned citizens participating in the ongoing air quality conversation.

The strangely daunting prospect of simply plugging in and connecting to a global community of concerned citizens able to solve an environmental problem as intractable as air pollution meant the Egg actually sat on our shelf for another week. Opening the kit seemed to be a ceremonial event for which we needed to be prepared.

So when, in early August, we set aside time to unbox and install the Egg, we made a wager as to how much time would pass before we were successfully gathering air-quality sensor data. Estimates spanned from having the device up and running within the afternoon to a few days. A more skeptical researcher estimated that it could take months, if ever, until sensor data were coursing through this plastic Egg and transforming into environmental solutions. You might find yourself engaged in such speculation as to what the final setup and output from your off-the-shelf sensors could be. This process is central to how citizen-sensing practices take form as ongoing contingencies of devices, environments, and engagement.

Once we had unpackaged the Air Quality Egg and scanned the different components of the kit, we next read through the seemingly straightforward how-to instructions. We began the setup by entering the device's serial number (or MAC address) into the Air Quality Egg Google Maps platform, where we also located and named our Egg.[59] With this, we were able to see the Citizen Sense Egg in South East London, situated within a wider global community of sensing citizens, albeit one that at the time numbered around 250 in population worldwide. Here was an apparently eager if niche community of Egg owners, ready to solve the problem of global air pollution. Once our Egg was on the map, we turned to setting up the device and posting data. You might find, as we did, that putting your sensor on a map is often the most basic and straightforward of the technical challenges you will encounter with the Egg.

The Air Quality Egg is formed of a pair of translucent white plastic Eggs: a "base" station Egg that at the time of testing posted data to the Xively platform, and a "remote" Nanode Egg that does the job of sensing air pollutants and gathering data via a shield outfitted with metal oxide sensors that detect nitrogen dioxide and carbon monoxide, along with temperature and humidity sensors attached to an Arduino microcontroller. Although there have been updates and a new version of the Egg since the time of this testing, with data now posted to a different platform and updates made in sensor setup, this was the device arrangement with which we worked at the time. As referred to in points 5 and 7 above, you'll find that the Air Quality Egg presents numerous dilemmas in the form of upgrades and fixes. These events are not unique to the Egg, because not only are these relatively new and unstable devices but also they inevitably succumb to the rapid rates of obsolescence that are characteristic of electronics.

During Egg setup, we next discovered that the power plug was configured for US electrics. Since we did not have a power converter, we had to set the project aside until we sourced an adapter for the UK context. This was a simple enough problem, but we found that it was just the beginning of several stages at which we realized additional kit would be needed to make the Egg function. Once we'd eventually powered the device, we found it did not perform the correct color sequence to indicate that it was collecting and posting data. Flashing color sequences were how we were to "gain awareness" of air pollution through the successful posting of environmental data to the platform. But the sequence of our flashing colors did not follow the same sequence outlined in the setup instructions. We spent some time trying to determine exactly what the color conversions were indicating, before discovering that the color issues were due to a bug that had been discovered several months earlier, in January 2013, but was still affecting devices shipped as late as ours, where the Eggs were no longer talking to the Xively platform for displaying data. An elaborate process then ensued of attempting to reprogram the Egg base and remote Nanode, following the official Air Quality Egg Google forum and FAQ instructions, which linked us to a seemingly straightforward video indicating how to fork a repository of code from GitHub to reprogram the Arduino microcontroller in the Air Quality Egg.

And yet, after completing the process of reprogramming, the Eggs were still not producing "data," neither in the form of flashing color nor in the form of line graphs on Xively or cryptic bar charts on the Google Maps Air Quality Egg page. Eventually, through multiple waves of turning the device on and off and reloading and verifying code, we managed to obtain blips of data, strange right angles in line graphs, and numbers apparently in parts per billion of nitrogen dioxide or carbon monoxide, but generally remaining inexplicable in terms of what measurement they presented, exactly. We had a modest assurance that the temperature and humidity readings might be somewhat correct, because these agreed with other sensors we had in operation in the same space, but converting the nitrogen dioxide and carbon monoxide measurements to a legible figure was a rather more difficult matter, since we did not have other verifiable sensors in operation for comparison. Our closest point of reference was with the official London Air Quality Network station, and here the measurements were not in units that could be easily cross-referenced.

By getting the Egg up and running, we had experienced what one of our developer–collaborators called the "'Hello World' of IoT." Getting an air-pollution sensor to post data was a basic achievement along the pathway of the IoT. Yet in the process of connecting the Egg we were more intent on asking this IoT "world" a host of other questions about how these devices were sensing air pollutants. What were these sensors sensing, exactly? How could we find out more about the hardware and software setup and the extent to which this could influence the data outputs? Precisely how did the "color-equals-awareness" engagement with environmental data work, especially when color did not signal pollution levels?[60] At the same time, what sorts of data were these, in terms of their accuracy, legibility, and legitimacy, when even getting a device to work, whatever the readings, seemed to be an achievement? And now that our device was operational, what were the capacities of the network of concerned citizens to which we were connected? From our use of the Air Quality Egg forums, it seemed that the communities to which we connected had more interest in hobby electronics and computer tinkering than mobilizing their data to influence air-quality policy or enforcement.[61] Indeed, the Egg was now circulating in the world in ways where maker communities effectively contributed free R&D by testing the device, while finding fixes and improvements through necessary troubleshooting.[62]

The universal citizen sensor, embedded in a community of global citizens, comes down to earth by working with the specificities of particular infrastructural configurations. The seamless plug-and-play logic and practices that such devices would promise continually meet with simple obstacles and more complex malfunctions. The imperative mood here might instruct you to plug in your Air Quality Egg and connect to other global and instrumental citizens. But your inability to complete the command or follow the instructions can multiply into a whole set of other practices, infrastructures, and relations that you will likely need to call on to fulfill seemingly straightforward instructions. Similarly, Lucy Suchman has pointed out that even a task as obvious as pushing a green button on a photocopier machine can give rise to confusion, adjustment practices, technical communities, and modified technological artifacts that materialize when simple button-pushing does not yield the expected results.[63] Instructions seem to guide a technical encounter, but they do not determine it. Instead, the "sense"

Figure 1.6. Testing different air-quality sensors and monitors, including a Shinyei particulate-matter sensor, during a walk in New Cross Gate, London. Photograph by Citizen Sense.

Figure 1.7. Setting up the Speck particulate-matter monitor in London, as part of the development of the Citizen Sense Toolkit. Photograph by Citizen Sense.

made of and through instructions materializes in the actual undertaking of a technical practice.[64] Practice-based research in this way is an approach that surfaces the many adjustments and deviations that can arise when working with technologies in lived situations. Instruments and instrumentalities frequently deviate from simple action and outcome. These are the misfires of the imperative mood, which generate the open-air instrumentalisms of citizen-sensing technologies.

Off-the-shelf sensors may promise that you can "simply connect"—and by extension also connect to greater air-pollution awareness and a community of global and instrumental citizens—but they often do not unfold in such a straightforward or liberatory fashion. Instead, they generate open-air instrumentalisms that deviate from the process of following instructions and setting up toolkits. Air-quality sensors do not always immediately function, either technically or in terms of their broader sociopolitical and environmental effects, but in the malfunction of devices and the reconstitution of instructions, other worlds in the making are generated along with instrumental citizens. The misfires that percolate through the imperative mood occur in part through anomalous technical arrangements that come into being, as detailed here, and in part through how these devices circulate and are taken up to address environmental problems. While these sensors proved to be anything but off-the-shelf, there is the possibility (if not the danger) that the promise of such modular and ready-to-use devices also could begin to inhabit the space of politics, encounters, and relations, for instance, in the form of off-the-shelf politics, off-the-shelf citizenship, and off-the-shelf public engagement. This is why it is important to ask what sorts of instrumentalisms are mobilized with apparently ready-made sensing technologies and sensing practices.

At the time of this writing, the Air Quality Egg has undergone many updates and is now available for sale in newer versions.[65] The website notes that "big improvements" have been made to the version 1 Egg that we tested. Indeed, its supplier, Wicked Device, no longer supports version 1, and Xively, the platform host, has terminated the data service that the version 1 Egg used. To post data, the Egg would need to be reconfigured to send to a different data platform. As the Wicked Device announcement states about this option, "That's a fair amount of work, and will require that you recompile and re-load your Egg with the new service destination."[66] The easier route is to purchase the version 2 Air Quality Egg, which guarantees an even more seamless plug-and-play experience. A new device replaces the defunct one, and the promise of technical action becoming democratic action is refreshed.

Around the time we had undertaken our own provisional setup with the Egg, I began to be asked by representatives from local governments, environmental

NGOs, and even air-quality officers in small nations whether they could replace their expensive monitoring instruments and networks with Eggs. My cautious reply was, not unless you'd like to spend considerable time dealing with misfires and miscalculations. While many of these off-the-shelf devices can be and are used in interesting ways that add to the scope of DIY practices, they are tetchy gadgets that produce a variable range of data that currently do not transfer well to the spaces of air-quality regulation. While a modest achievement can be made in getting a sensor device such as an Egg up and running, its flickering displays and data outputs do not necessarily sync well with the expanded technical, social, political, and environmental requirements of air-quality governance in its usual sense. For this, you might need to engage with even more versions of the how-to, including points 7 to 10 given earlier, which indicate how the technoscientific configuration of an air-quality sensor and the data it generates depend upon extended infrastructures to make sense.[67] These are infrastructures not just of technical capability but also of stabilizing data-as-evidence to address the experience and event of air pollution.

Whereas the Air Quality Egg seems to promise that air-quality monitoring can become a relatively effortless affair, many plug-and-play sensors require considerable effort to become operational. Updated and upgraded versions will still require ongoing maintenance and fixes, as well as skilling up to learn about technical configurations. At the same time, air-quality sensors are now proliferating apace, with many more sensors becoming workable as plug-and-play devices. The PurpleAir sensor and IQAir, for instance, are now in regular use to monitor air pollution from wildfires, traffic, industry, and more. Yet every sensor still raises questions about the verifiability of the data these plug-and-play devices generate, as well as the protocols and practices used during installation. While in no way meant to deter you from testing out citizen-sensing technologies, this how-to setup that involves working across standard instructions as well as actual practices undertaken is meant to demonstrate the misfires and miscalculations that proliferate when inhabiting the imperative mood and when working with the genre of the how-to and the toolkit.

The Reluctant Prototype

Parallel to, and perhaps even in advance of, working with the Air Quality Egg, we were in the process of developing our own prototype air-quality-sensing toolkits, which I detail here as the third example of working with sensors in practice. To begin, we assembled provisional groupings of nitrogen dioxide and particulate-matter sensors that we had used in a pilot walk in the New Cross area of London in early July of that same year.[68] You might find that as you progress from

Figure 1.8. Investigating different sites, including riverfront and roadside locations, for possible air-pollution emissions during a walk in New Cross Gate, London. Photographs by Citizen Sense.

following O'Reilly and Instructable tutorials, your own devices assemble neither as makerly stuff nor as off-the-shelf kit, but as particular prototypes that are cobbled together in a cut-and-paste and makeshift way. A sensor configuration that works in one setup can be morphed over to another expanded kit, and code passes along on these various iterations or is drawn from libraries to create a new workable concoction.

In just this way, we were attempting to put together a possible prototype citizen-sensing kit that we could use while undertaking fieldwork in the United States, where we were researching fracking-related pollution in Pennsylvania.[69] Multiple citizen-sensing activities to address pollution and public-health concerns were already underway in Pennsylvania. In this context, we wondered what role a prototype kit could play in engaging people to ask questions about environmental data, how the data are generated, and their effects in addressing the problem of air pollution. How might data be shared and collectivized? How might data travel differently from the current, if complex, modes of reporting on well locations and pollution levels? In the process of building a kit comprising multiple air-quality sensors, including nitrogen dioxide, carbon monoxide, temperature, humidity, and particulate matter 2.5 ($PM_{2.5}$) sensors, we found that as many questions were raised about the validity of data that might be generated from such a kit as the possible promises and expectations that could be raised by circulating environmental technologies to communities affected by pollution from fracking. We were beginning to engage with sensors in the open air, not just by moving them to actual sites of pollution detection, but also by collaboratively testing them with communities knowledgeable about documenting pollution through environmental data collection and analysis. In this context, open-air instrumentalisms and instrumental citizens multiplied and abounded even further.

This toolkit in the making demanded that we think through the instructions and how-to pointers that we might provide to make the kit legible and usable. Numerous questions came up when we thought about how these kits might be used in the field: Would there be a manual with instructions for use? Would each sensor be explained in relation to what it senses and how chemical detection optimally works? How long would the sensors need to operate to collect usable data? How long should the sample rates and duration of monitoring be? Would the sensors work only if stationary, and should instructions be given to keep the kit stable during use? Would data be made available to individual participants, or would it be shared collectively on a web platform? Should data be given locational information or be made anonymous? What instructions might participants need to analyze the data in order for them to be meaningful and actionable? Would the sensor housing skew the readings in any way? What are the base readings of the sensors, and are we sure they are properly calibrated? Would the sensors or pollutants interfere with each other? Could we be certain they are sensing exactly what they are meant to sense? Finally, would the kit be damaged in shipping from London to Pennsylvania, and what adjustments might need to be made in the field?

These questions connected up with our attempts to ensure that multiple participants' engagement with the kit might be collaborative and experimental from the beginning and not only a functional end application. At this point in the development of sensors, we queried the notion that by collecting data—accurate, skewed, or otherwise—environmental politics would be more readily democratized or facilitated. We sought to work through this seemingly more instrumental–functionalist agenda by following and querying the possible trajectories of data to action. Yet while we sought to critically examine the role of environmental monitoring technologies in forming practices and politics, rather than simply becoming advocates of this approach, we also had to take seriously the instrumental logics of these devices and the citizen-sensing practices they activated and organized. These reworkings of instrumentality became part of how we experimented with making alternative citizen-sensing toolkits that could engage with the practices and concerns of participants engaged in preexisting monitoring projects and generate usable data, while also opening into other engagements with environmental problems.

So, with all of this in mind, in fall 2013 we began making prototypes to test how sensors generate, influence, and operationalize environmental data. Our version 1 Citizen Sense Toolkit initially consisted of two primary devices: a sensor shield pulled from the Air Quality Egg, which included nitrogen dioxide, carbon monoxide, temperature, and humidity sensors that we attached to a Grove Board to add a real-time clock; and a Raspberry Pi microcontroller. The version 2 Citizen Sense Toolkit comprised stand-alone sensors (rather than a pluggable sensor shield), including carbon monoxide, nitrogen dioxide, $PM_{2.5}$, temperature, and humidity sensors; we added our own resistance configurations, which we found considerably improved the readings in comparison to the Egg shield.

With both of the preliminary Citizen Sense Toolkits, the first intention was to get the devices up and running, because in the process of making the kit even more questions had emerged about the how of the how-to. The second intention was to disassemble and reverse-engineer more black-boxed technologies, such as the Egg, which on one level required all sorts of capacities and resources to function and on another level had rather unclear information about how the hardware and software were put together, how the sensors were configured, what resistance was used and how this affected data outputs, and how continual changes of the data platform "back end" (from Pachube to Cosm to Xively) could affect the data's form and analysis. You might find yourself asking similar questions if you attempt (or have attempted) to make sensor toolkits. Repurposing, retooling, and reworking become key techniques in assembling and questioning the forms of action that these toolkits generate.

In this way, the Citizen Sense Toolkits were built through information from multiple forums, since there was no single official forum from which we might obtain guidance on how to make the monitoring technology "work." We developed even more iterations of the Citizen Sense Toolkit, including a version 3, which included a Speck particulate matter sensor, an analog BTEX (or benzene, toluene, ethylbenzene, and xylene badge), a Frackbox (for monitoring volatile organic compounds and nitrogen oxides), a monitoring logbook, and a data platform. This version of the Citizen Sense Toolkit was later used to monitor pollutants from fracking at thirty community locations from fall 2014 to early summer 2015.[70] Along the way, we foraged for diagrams and work-arounds, forked code from GitHub, and shoveled piles of breadboards, resistors, and cables across desktops. By rebuilding kits and gaining another perspective on the hardware and sensor configurations, we were also able to observe along the way what technical resources, capacities, and infrastructures these technologies require, as well as the decisions that were made or elided to create the monitoring kits in these particular ways, and the domains inhabited to generate and circulate environmental data through these contraptions.

By working with toolkits for sensing air pollution, we tested how these technologies enable certain types of monitoring and generate questions about the limits and possibilities of each of these monitoring practices for addressing environmental and political problems. The practice of making prototypes and setting up off-the-shelf sensors becomes a way to work through the instructions, promises, functions, and malfunctions of these devices. It also generates openair instrumentalisms. In the process of procuring guides for air-quality sensors, making kits, following instructions, and installing devices in the open air, a number of splintering pathways came into view by deviating from a straightforward approach to these devices. Online forums read as tales of ongoing struggles to set up sensors, to maintain their operations, and to update and adjust when upgrades are available. FAQ sections are brimming with queries about connections, data, and modifications. Platforms bear the traces of half-finished efforts in running sensors, where maps of monitoring locations click out to nonexistent line graphs or inexplicable charts. These open-air instrumentalisms began to take on a more-than-technical quality as sensors were readied for installation and use, where an initial success at connection splintered into multiple concerns about the use or relevance of these devices.

But as noted earlier, this is not to say that devices never arrive at a condition of organized use or implementation. Instead, it is to signal how setting up citizen-sensing technologies is an ongoing trial, a back-and-forth effort of testing and tweaking. At the same time, despite the democratic selling points, many devices remain tied to practices focused primarily on technology and "making" and so can become somewhat self-referential in their pursuits, thereby missing the promise to address—and even improve—environmental problems. Yet if, as Dewey has suggested, the "invention of new agencies and instruments create[s] new ends," then how do these new instruments "create new consequences" and "stir" us to "form new purposes"?[71] This is a question about instruments and instrumentality, which the next section considers in a more reflective key.

HOW TO DEVISE INSTRUMENTS

When Whitehead asserts that "every science must devise its own instruments,"[72] he is referring in part to the need for specific tools to be formed in relation to modes of inquiry. A study of ecology materializes as an inquiry that significantly differs from a study of philosophy. Here Whitehead notes, "The tool required for philosophy is language."[73] His statement has a multidirectional character. It suggests that tools are required for distinct scientific practices, formed through devising and using distinct instruments. This assertion could point to physics and mathematics, biology and atmospheric chemistry, in addition to philosophy. It might also indicate how the citizen-sensing practices form with and through distinct instruments and instrumental processes. Yet are instruments also defining entities for these practices, which might variously be characterized as what Ruha Benjamin has called a "people's science"?[74] And if so, how are these instruments further characterized by modes of practice and not just their distinct form as tools? In other words, citizen science and citizen sensing cross the spectrum of possible tools and subjects of inquiry. Yet it is not just the actual instrument used that is the defining characteristic, but also the mode of engagement and relationality set in motion that remakes instruments, scientific practice, and inquiry. This section considers how instruments materialize along with practices of inquiry and inquiring subjects—the instrumental citizens who would undertake sensing projects.

Instruments do work in the world. They can make undetectable phenomena evident. They tune in to other registers of experience, and generate perceptive practices that remake sensory worlds. A list of instruments devised along with scientific techniques could extend to epic proportions, spanning the fantastic and the precise. If the air pump has featured to demonstrate the emergence of a particular mode of objective science,[75] it has also been the source of much attention in producing universalized subjects who are seemingly detached from making their objects of inquiry and knowledge.[76] In this way, instruments and machines have served as devices for differentiating the contours of a rational human subject from an automaton or a duck.[77] Conversely, technical devices, such as engines, are not mere instruments but are generative of new subjects, milieus, and relations.[78] Instruments might also seem to be something distinct from the contours of the human body. Still, as writings on the cyborg have demonstrated, instruments can remake technologies, subjects, bodies, relations, environments, and politics, as well as what counts as scientific inquiry.[79]

Environmental sensors as they are used within citizen-sensing practices are similarly wide ranging. Hygrometers and anemometers, barometers and thermometers, as well as metal oxide and electrochemical sensors for detecting air pollutants: there is a roving toolkit of borrowed, appropriated, hacked, and repurposed parts with which citizens work to attempt to document environmental disturbance. Here are multiple instruments with different tunings, standards of measurement, modes of observation, political effects, and world-making capacities. If instruments are integral to the practice and definition of what counts as science, then you might wonder how citizen science and citizen sensing, with their DIY and makeshift instruments, begin to challenge and rework not just instruments but also what counts as science. What are the instruments of citizen sensing? How would citizen sensing vary in relation to those sensors listed earlier, from PuffTrones to the Air Quality Egg? How do these devices contribute to the formation of diverse practices of inquiry? What are their capacities for transmogrifying the evident to make new forms of evidence? These are questions to ask along the way while wondering about these instruments in the making.

Instruments are a long-standing topic of investigation in science and technology studies.[80] Rather than tracing the historical lineages and social–epistemic formations of instruments, however, a how-to approach charts the uneven and sprawling ways in which contemporary citizen-sensing instruments help pursue environmental and political agendas and how they at times fail to realize these outcomes. This is a way of working within while also reworking instruments toward open-air instrumentalisms. The instruments of citizen sensing demonstrate how apparently instrumentalist versions of evidence-based politics can give rise to diverse and inventive citizen-based and collective practices through the very attempt to gain influence by collecting data. These practices complicate an easy critique or adoption of instrumentalism. They also reinvent relations with instruments and instrumentality. At the same time, instruments or tools are already mutually constituted with practices so that new citizens and worlds con-cretize through engagement with instruments, but not as a linear process.

Instruments are invariably involved with social relations. Any change to them, as Bruno Latour has suggested, will also shift social conditions. As he writes, "Change the instruments, and you will change the entire social theory that goes with them." Here Latour is engaging with the work of Gabriel Tarde to note how "science is *in* and *of* the world it studies," where instruments become crucial to social relations as they are performed, lived, and understood.[81] A change of instruments and the standardization of instrumental processes also "in-forms" the worlds sustained and set in motion.[82] Instruments can operationalize effects. Such effects are less likely to materialize through an instrumental script and more likely to concretize through the social worlds and political subjects that assemble along with instrumental processes. You might wonder if there is also a how-to aspect to Latour's assessment of Tarde. In other words, how do you change the instruments so that you can also change the entire social theory that goes with them? In its search to devise instruments, the how-to guide could be a call to undertake experimental engagements that generate ways of working with and through new technical arrangements, infrastructures, and modes of governance.[83] The how-to guide is not simply the study of a technical problem; it is also an encounter with the potential of other social worlds.[84] When thinking about how to devise instruments, you might consider how changing instruments also changes the possibilities of encounter, engagement, and relation.

Instruments and Instrumentality

Instruments are the tools, devices, and contraptions constituted as they do work in the world. An instrument can be a sensor, a data logger, a toolkit. There are also conceptual instruments, discursive instruments, and policy instruments. An instrument could standardize and measure and also construct and generate. Instruments and instrumentality mobilize inquiry and experimentation while organizing observation and action. Are instruments generative of an expanded instrumentality, or are they prescriptive in their engagements and outcomes?

Instruments are often described as "mere" or "passive" or "functional" de-vices. Simondon suggests that an instrument-based view of technology tends to be reductive. He writes that the technological object has been "treated as an instrument" considered in relation to economics, work, or consumption but has not been engaged with through philosophical or cultural deliberations.[85] Unlike cultural objects, he suggests, technical objects are relegated to "a utility function" and do not enjoy "citizenship in a world of significations."[86] For Simon-don, the designation of a technical object as an instrument is a way of focusing on its functionality only, where instrumentality seemingly has a predetermined outcome: to complete the task at hand. By suggesting that this is a way of denying technical objects a sort of citizenship, he seeks to diversify the entities from and through which meaning and sense—meaning as sense—materialize. Tools and technics, in other words, are cultural relations and expressions.

However, while in Simondon's analysis an instrument might be seemingly fixed in its capacities and modes of observation or operation as well as outcomes, it is also subject to retooling. As Whitehead notes, language is not simply a tool used by philosophy; instead, "philosophy redesigns language in the same way that, in a physical science, pre-existing appliances are redesigned."[87] This redesign occurs in part because of the breakdown of that instrument, which occurs at the edges "of expressing in explicit form the larger generalities—the very generalities which metaphysics seeks to express."[88] As instruments are engaged in processes of inquiry, they are worked and reworked toward the edges of inquiry. These edges of inquiry can be as sociopolitical as they are "technical." Drone pilots and protectors who push technology to its limits convey this, as discussed in the introduction to this book. But breakdown and retooling are not the only conditions of this instrumental engagement. These conditions occur because instruments—in this case, language—are searching toward propositions of fact that are also referring to the pluriverse or worlds needed to sustain those facts.[89]

Flat-pack cosmologies surface again here, but from another angle. A world is not ready-made from a toolkit, nor is a toolkit as ready-made as it might have seemed to be. Instead, a world is required for instruments to be put to work, making both tools and worlds somewhat indeterminate in the inquiry to be undertaken, because they are in process. An instrument might reach toward something more fixed and absolute, since it will require its world to make sense, yet these are both in the making within practices of inquiry.[90] The instruments and instrumentality that might have seemed to project toward a certain outcome instead generate open-air instrumentalisms. They form through relations and deviate from a fixed purpose. They take shape through distinct modes of inquiry.

These many components of scientific practice involve what Jenny Reardon and her coauthors refer to as the "material relationships that are part of knowledge-making practices, including political, social and cultural ones."[91] Instruments are always connected to a "multiplicity of entangled apparatuses" that include ethics and justice.[92] While the focus could easily lead to human-makers taking up instrument-toolkits to address environmental problems, such a perspective would further demonstrate how instruments, observations, observers, and phenomena are entangled such that world making is a distributed and multi-agential affair.

In this sense, apparatuses for Barad "are constituted through particular practices that are perpetually open to rearrangements, rearticulations, and other reworkings."[93] Resurfacing here is a certain breakdown and redesign—or retooling—that occurs not as the work of a willful human subject but rather as part of the shifting conditions in which instruments unfold through instrumental operations and relations. Open-air instrumentalisms are multi-agential and not only are the work of makers or tools but also erupt through situations, practices, other entities and relations, along with attempts to build breathable worlds.

In the process of making instruments, you might wonder whether your approach to sensing environments has become somewhat "instrumental" or, as usually designated, overly functional. But as this discussion begins to suggest, even that which seems to be defined as an instrument, and its instrumental outcomes, begins to break down and be retooled through practices of inquiry. Although instrumentality has acquired a negative connotation—to say that something is "instrumental" is to suggest that it is a grossly efficient means to an end—these critiques of a certain mode of causality deserve another look in the context of working with citizen-sensing instruments.[94] Although citizen-sensing technologies are often wrapped in the promise of a simple means–end practice, where sensing the environment will generate political change, the instrumental operations of these instruments are never as simple as this. Instrumentality can demonstrate other modes of effect and effectiveness, not that of a reductive cause and effect but rather a multi-agential making of worlds.[95] Although instrumentality might seem to generate a limited set of engagements, this revisiting of instrumentality from within the milieus of instruments-in-practice shows how more expansive modalities of action and practice can materialize.

Instrumentality is a mode of experience that might be productive of particular observations, expressions of citizenship, and relations with other collective entities for acting on problems of environmental pollution and environmental harm. Instrumentality, in this sense, necessarily becomes experimental in the process of undertaking concrete action. "Instrumental experimentalism" was a term and concept that Dewey used in a somewhat interchangeable way with "pragmatism" to refer to the contingency of "ends" within a philosophical— or democratic—project.[96] On one level, Dewey was accounting for the rational unfolding of concepts in concrete situations that is central to the pragmatists' approach to the instrumental. On another level, Dewey indicated how instrumentalism had implications for democracy—as a conceptual project always likely to generate struggle,[97] contested relations, and modes of governance that are not direct or effortless instantiations of democratic principles.[98] Or as West has suggested, such a "future-oriented instrumentalism," which ran the risk of heroic or individual approaches to creative democracy, was also a search for strategies of "more effective action."[99]

A propositional end might serve as a guide for concrete action, but it is always provisional and inevitably reworked through concrete experience and practice. Because an end is not merely arrived at, moreover, it is in many ways radical in relation to the instrumental experimentalisms it operationalizes but from which it also deviates. Instrumentalism for Dewey is about a process of experimentation, inquiry, and discovery. In this sense, it would be possible to say that instrumentalism has always been experimental. From instruments as logical concepts to instruments as material technologies, the rational unfolding that would occur instead gives way to even more prospective instrumentalities.

In this respect (and in contrast to Simondon), the instruments of citizen sensing are not instrumental enough, since they seem to guarantee an outcome that would foreclose the very undertaking of citizen-sensing practices as concrete experiences. This *a priori* designation of an outcome reduces not just the instruments and instrumentality but also the instrumental citizens that would materialize through these practices. Although the terminology is different, with this Simondon might agree: the conventional promises of citizen sensing constrict instruments into functional outcomes, a process that forecloses inquiry or experimentation. Here Simondon might be inclined to admit such instrumentalism to his analytical toolkit, because this does the work of reclaiming the processes of inquiry and open-endedness that he suggested were more appropriate to understanding and transforming human relations with technology. Shutting down and narrowing inquiry, as Dewey suggests, limits the modes of experience and political engagement that could be possible.

The instrumentalism developed here takes a cue from these pragmatist approaches to experimentation and inquiry and is informed by the open air that James found was necessary to practices of inquiry. "A pragmatist," as James writes, "turns away from abstraction and insufficiency, from verbal solutions, from bad *a priori* reasons, from fixed principles, closed systems, and pretended absolutes and origins." Instead, pragmatism involves a turn "toward concreteness and adequacy, towards facts, towards action and towards power."[100] Such an orientation directs inquiry toward "the open air and possibilities of nature, as against dogma, artificiality, and the pretence of finality in truth."[101] *Open air*, as I develop the concept with and beyond James, refers to an operationalized and prospective approach to inquiry. *Open air* pertains to lived experience, to processes of inquiry as they are unfolding, rather than to doctrines to which inquiry is made to conform.

Expanding on this aspect of James's work, Dewey suggests that instruments in the form of ideas become "true *instrumentally*" through how they "work." The working aspects of ideas were far more relevant than the final outcome that might seem to offer up a resolution or promise of truth. At the same time, instruments are also "a program for more work, particularly as an indication of how existing realities may be *changed*."[102] Open-air instruments and instrumentalisms are toolkits for action; they are able to generate change, above and beyond a static pronouncement of truth. Putting instruments to work in the open air is a process that is integral to practices to make more breathable worlds.[103]

Expanding on James and Dewey, I move from the unfolding of logical instruments to practices with technical instruments to suggest that instruments such as citizen-sensing technologies are more than a means to an end. As it turns out, it is only by undertaking practices and engagements with and through instruments that contingent relations and capacities—as well as citizens and worlds—begin to materialize, demonstrating that instrumentality has never been quite so straightforward as it might have seemed.[104] This is the scope of open-air instrumentalisms: to demonstrate how sociotechnical practices are set to work and how they potentially work toward more breathable worlds. While Dewey sought to clear up the confusion about the terminology and meaning of instrumentalism, I work with this productive dissonance to query the trajectories and outcomes of sensing instruments. I propose the term *open-air instrumentalisms* as a way to indicate this revisiting and reworking of instrumentality within the context of DIY environmental sensors. This concept and term is about more than logical propositions, since it also expresses the prospective qualities of instruments and instrumentalities. Open-air instrumentalisms are a strategy and practice for breathable worlds, where breathability is an approach to instruments that opens into processes and exchanges.

The open air punctures any closed logic of instrumentalisms. Despite the imperative mood of guidebooks and toolkits that would suggest a quick passage from flat-pack cosmologies to actionable gadget, you will find that there is not a simple way to bend technology to your will. Instead, here are the toolkit, the instructional, the guidebook, and the instrument unfolding into the open air of instrumental experimentalism. Rather than instrumental reason giving way to a singular means–end trajectory, these are open-air instrumentalisms that, when put to work in the world as practice, concrete experience, and contingency, engage with and generate multiple inhabitations that struggle toward breathable worlds. Instruments not only contribute to organizing inquiry in particular ways; they also distribute inquiry across multiple entities and relations, creating new communities of inquiry—something I will address in the chapters that follow. Instruments are involved in tuning and in-forming environments, worlds, and political subjects that further trouble the usual scope of instrumentality: those instrumental citizens.

Instrumental Citizens

By drawing on multiple and diverging thinkers, I expand on the notion of what instruments and instrumentality might mean or generate. There are many different uses of the term *instrument* across these thinkers, and they are by no means synonymous. The instruments of Simondon are merely functional technologies; the instruments of James and Dewey are theories and ideas put to work in the world; the instruments of Whitehead become part of practices of inquiry; and the instruments of Barad expand out to relational, material, and entangled apparatuses. If Whitehead's remark quoted at the beginning of this section has much to say about science and instruments, it says less about the subjects caught up in these instrumental practices. Who or what are these instrumental entities? If "citizens" are monitoring environments with sensing instruments, do they become instrumental citizens? Are they instrumentalized in the conventional or in the pragmatist sense? And do they realize new political competencies through their instrumentalist practices, which theorists of feminist technoscience, Black studies, and Indigenous theory develop as strategies of retooling technologies and action?

Suppose the citizen took shape through the pragmatist sense of the instrumental. In that case, it would mean that the democratic commitments of political subjects are continually put to work. Through this work, the very meaning of *citizen* would come to have consistency. Citizen practices then generate the reality and community of citizens, as well as the transformed instruments that would further spur this work along.[105] Instrumental citizens are not rationalized actors completing a designated task—the reductive or functional sense of *instrumental*. Rather, they are contingent subjects involved in making and remaking—tooling and retooling—political life. As ongoing work in environmental justice has demonstrated, the retooling of instruments occurs along with the transformation of politics and action to work toward less polluting environments.[106]

A citizen-sensing kit comprises citizens as much as sensors. Yet the "citizen" is not an entity that can be wired and coded in the same way as a microcontroller. Instead, what the citizen is or could become is in-formed by sensors and the extended milieus in which they operate. The "citizen" in these toolkits is meant to be an action-based entity. This is a citizen imagined to be an empowered and effective technophile. The instrumented citizen is an instrumental citizen, in the usual sense of realizing a stated outcome through direct and efficient action. In this sense, the instrumental citizen is a tool-kitted citizen.

Yet the logic and expediency of sensing instruments and instrumental citizens, along with their ability to effect change, is a narrative that leaves the details of technical and political engagement unexamined. It is worth pausing to examine in more detail the usual diagram of how action is meant to unfold and how the designations and expressions of citizens and citizenship are performed through sensing technologies. For instance, Plume Labs, which has developed a wearable Flow sensor and an AI-powered app for forecasting air pollution, focuses on how citizens as sensors might monitor their own air to protect themselves and their families from high pollution levels.[107] While they suggest that collective sensing projects are possible, Plume manages and oversees the collected and collective data so that they are not readily available for use and analysis by communities. Plume emphasizes personal action and protection and brackets collective action as a less relevant or feasible use of sensors. This arrangement might assure user–consumers that by monitoring their air they are not sliding into the dangerous depths of citizen activism. Instead, they can maintain a more innocuous engagement with technology to protect themselves and their families. Engagement becomes a more nuclear and normative undertaking, less inclined to the sprawling affiliations of democratic communities beyond the family.[108]

Such a configuration of subjects is one of vulnerable and responsible family members managing their personal air space in a politically neutral manner. The air here becomes more like an atmosphere of air-conditioning and security—an area of instrumental control rather than open-air instrumentalisms.[109] Citizen-sensing instruments facilitate these environmental practices, thereby shoring up a particular citizen-as-consumer engagement with the problem of air pollution. This is not an isolated example of how many consumer-based air-pollution technologies are now being promoted, whether in the European Union or the United States, China or India: the focus is on managing and protecting oneself and one's family members in controlled personal spaces. Instruments and instrumentality give rise to awareness, personal protection, and responsibilization, rather than collective organization, environmental transformation, or strategies for building breathable worlds.

In other words, these "aware" subjects are not directed to intervene in current operating conditions to undertake democratic struggles toward more breathable collective atmospheres. Instead, they are made aware so as to better manage their own individual exposure. Of note here is that citizen-sensing technologies for monitoring air pollution are increasingly shifting away from DIY and makerly technologies toward finished consumer products. As a result, instruments could become locked into organizing and directing technopolitical engagements. In this context, they do not as readily give rise to open-air instrumentalisms. Instead, they potentially direct user–consumers to a series of corrective or adaptive actions not dissimilar to a cybernetic logic that Simondon critiqued for its functional approach to technical objects, which overlooked how technologies undergo processes of concretization as they in-form subjects and environments, or citizens and worlds.[110]

Here the citizen also becomes utility-like in the imaginings of citizen-sensing technologies, as an entity able to singularly and instrumentally effect change. But the reflections of Dewey suggest that we might consider other forms of instrumentality in relation to politics. Following the pragmatists, Antonia Majaca and Luciana Parisi also suggest that instrumentality is not instrumental, at least not in the way it is usually conceived. Through a more thorough engagement with the logic of *technē*, they suggest it might be possible to try "reversing the very understanding of instrumentality," which could be undertaken "by fully acknowledging instrumentality, politicizing it, and ultimately transcending it."[111] In their estimation, transcending instrumentality entails recognizing that subjects are also contingent, and this contingency is where the political materializes through concrete practices.[112]

Instrumental citizens might be sparked through instrumental practices, but they are in formation since they are involved in making worlds. In any given situation, instrumentality involves prospective engagements, which constitute the political. A further elaboration upon the concept of an instrumental citizen would then involve taking up Dewey's notion of "instrumental experimentalism" as the mobilization of what a political subject is or could be.[113] As previously discussed, this is less a fixed mode of engagement and more an opening into how the citizen as attractor and force can stir people to new purposes.

Such engagements do not constitute technical solutions but are provisional practices that transform technologies and politics. Instrumentalism is not a test for the sake of a test or an experiment for the sake of an experiment. Instead, it is a practice guided by ideas, technologies, and toolkits that seek to do a certain amount of work, and even possibly (political) transformation, in the world. The fact that instruments and instrumentalities are unlikely to fulfill their stated aim is not a limit but rather is crucial to testing instruments and ideas for further development. As James writes when discussing the work of Dewey, "*theories thus become instruments, not answers to enigmas, in which we can rest.*"[114] This may explain why Dewey used the term *instrumentalism* as well as *instrumental experimentalism* to describe this putting to work of instruments.[115] *Experimental* describes the contingent and open-ended modes of action, but instruments are the things and concepts put to work and reconfigured through experimental processes.[116] Hence the relevance of this discussion for understanding what a citizen—as an instrumental citizen—is and might become through engagements in the open air.

The point of revisiting and reworking instrumentalism as open-air instrumentalisms is not to recuperate a reductive notion of technical or political action but instead to consider how neither citizens nor machines have ever been so instrumental in the usual sense of the word. The adoption of a citizen-sensing instrument does not make for a more direct realization of a citizen-scientific or citizen-political impact. Instead, it organizes modes of inquiry, social relations, facts, and worlds. Instrumentality describes the instrumental commitments that unfold through practice and in the open air. The logic of sensing used to promote these technologies as a direct solution to environmental pollution could be understood as a form of instrumental reason that diminishes a more contingent and experimental understanding of instruments and inquiry. Instrumental reason is bound to bend, and instruments unfold through contingent operations. Political projects and struggles thus become activated and entangled with instrumental experimentalism.

When taken up and put to work, the instruments and instrumentalities of citizen-sensing technologies break down, open up, and are retooled through particular practices and situations. Instrumental citizens, in this sense, are political subjects (which are also not necessarily always human subjects) that work through the problem of sensing environmental pollution to make more breathable worlds. Here it might be possible to suggest that technical objects could be granted "citizenship in a world of significations," as has previously been discussed through Simondon. However, such citizenship is never settled. Instead, it is undertaken through the differential and multidirectional practices of human and nonhuman instrumental citizens as they sense and rework toolkits toward more breathable worlds.

HOW TO BUILD NETWORKS

Citizen-sensing technologies delineate modes of action—including making and coding, monitoring and data collection—that become instrumental yet prospective engagements. The commands to "get practical" as well as the exhortation "enough discussion—it's time to build!" are calls to action that toolkits and guidebooks, as well as the how-to and the imperative mood, organize and deliver. Yet these instrumental imperatives take on a very different meaning once you rework and retool instruments toward open-air instrumentalisms.

Figure 1.9. An example of citizen-sensing activities in northeastern Pennsylvania, where a resident installed a CCTV camera to document industry traffic and activity at the property perimeter. Photograph by Citizen Sense.

By building and getting "practical," a shift in current operating conditions is meant to occur. Such instruments and instrumentalities demonstrate how technoscientific practices, instruments, subjects, and worlds are collectively generated, along with an estimation of what the consequences of these instrument-worlds might be. The "practical" is what James refers to as "the distinctively concrete, the individual, the particular and effective as opposed to the abstract, general and inert."[117] Yet for James, when expanding from pragmatism to radical empiricism, the distinctively concrete refers not merely to things but also to relations.[118] Or as Haraway has suggested in her discussion of yet another instrument, the air pump, "nothing comes without its world."[119] You might find that "getting practical" requires a greater engagement with the sprawling relations, networks, and worlds that materialize along with instruments.

Constructing toolkits and connecting sensors are practices that further expand into techniques for building networks and worlds. Getting practical is always an encounter with and formation of relations. The setting up of one device moves from making or plugging in a sensor and piping data to a platform to connecting with and comparing data across multiple sensor nodes. But this computational approach to networks is only one way of configuring what a network is or might be as it concretizes through citizen-sensing technologies. While it might at first have seemed the primary focus, when taken into the open air, a sensor becomes one small component within a broader project of addressing environmental pollution. Indeed, when it comes to monitoring air quality, communities are often already mobilized in various ways to document and address pollution. Networks are in the making, but they do not start from zero. Sensors become part of community-organizing practices, and technical relations transform in the process.[120]

Because networks are already at work in the world, Citizen Sense set out to learn from and alongside existing environmental monitoring practices. During our research, which involved conducting online searches, attending community meetings, arranging interviews, making site visits, distributing logbooks, hosting mapping workshops, and guiding monitoring walks, we found that communities were monitoring air, water, noise, and traffic by using analog and digital sensors, including particulate-matter sensors, air-pollution badges, decibel meters, FLIR cameras, video and photography, and CCTV installations. Communities also used professional lab-testing services, gathering and consulting planning documents, keeping track of changing land surveys, monitoring policy and regulation, and petitioning for changes and improvements to environmental controls.

Practices of identifying pollution sources and using tools to monitor emissions then become just one aspect of different modes of inquiry and action. Communities work with existing networks for organizing environmental projects, and they find ways to contribute to and build on these to address concerns about environmental pollution. They also contact regulators and policy makers to register complaints about pollution, host community meetings, gather evidence about health conditions, give public testimony, share news on social media, set up teleconferences, contact experts and public figures to extend and amplify networks, and document pollution with assorted sensing technologies. Instrumentalities shift here. "Building" something involves much more than making a digital device operational. A project to monitor and address air pollution involves building community-monitoring networks as ongoing, iterative, and contingent practices that make and maintain technical, social, political, and environmental infrastructures.

Perhaps somewhat different from citizen science, citizen sensing has a more specific focus on digital toolkits and devices. The organizational, collective, and environmental aspects of monitoring might initially seem to be of lesser importance. However, in this way of configuring what a community-monitoring network is or might become, it is clear that sensor toolkits develop into much more than digital gadgets or makerly components. By working with situated environmental problems, citizen-sensing practices and technologies quickly become bound up with wider networks of environments, communities, institutions, and politics. The accuracy of monitoring devices, the monitoring protocols used, the legitimacy of the data, and the agendas of residents and communities all come into play as factors influencing the techniques of environmental monitoring and the data gathered. Citizen-sensing practices move from the more reductive diagram put forward by the Air Quality Egg to shift instead into distinct networks of inquiry and political contestation. In the process of making sensors, you might find that these technologies proliferate along with different networks that include the communities of inquiry that make, install, query, and operationalize citizen-sensing technologies.

Communities of Inquiry

The practice of building a community-monitoring network involves building and drawing on communities of inquiry. The process of taking an instrument into the open air does not merely consist of testing or setting up a device. Instead, a toolkit develops along with networks and inquiries. "Community of inquiry" is a concept that Charles Sanders Peirce developed to describe scientific modes of inquiry and how reality, facts, and truth are settled on through collective processes.[121] This phrase was taken up by other pragmatists, such as Dewey, to describe how concrete practices of inquiry generate realities that are particular to groups undertaking such work. For Dewey, these modes of inquiry become political, informing the possibilities and struggles of democratic life.[122]

The how-to can involve multiple processes of inquiry. But this is not merely an abstract set of instructions followed by a universal subject. Instead, the how-to as inquiry is situated within communities. Together with these formations of communities, inquiry, and facts, instruments transform with communities of inquiry.[123] This expands communities of inquiry to include nonhumans in their technical and fleshy arrangements and instantiations. This approach to the how-to process becomes less about a maker tinkering with a digital object and more about the collective constitution of worlds. How-to is a way of organizing and asking how to go about something, including how to make a world. How are communities of inquiry organized in relation to environmental problems? What are their practices, tactics, and strategies? But the question does not merely document the occurrence of networks. Instead, it also contributes to the prospective formation of networks. This is part of what you might attend to when making a citizen-sensing toolkit.

As a prospective undertaking, inquiry is a mode of transformation. Another approach to the how-to materializes here, where "how" indicates or asks in what manner, by which means, and how it might be possible to organize ways of life. "How" indicates procedure, practice, and process. The imperative mood shifts to become less commanding and more aligned with a specific obligation and necessity. In *Imperatives to Re-imagine the Planet*, Gayatri Chakravorty Spivak outlines a mode of the imperative that involves recasting the relations of subjects through planetary connections that exceed that which can be designated or made commensurate with subjects.[124] The imperative in this sense is as much an opening as a responsibility, a proposition as well as a commitment to justice. "How" attends to the mode of engagement and the imperative of attending to what is at stake in attending to and attempting to address (if not redress) planetary troubles.

Here, the how-to opens up to engage with another register of the imperative: the crucial actions that contribute to lived engagements that remake worlds. Elaborating on this aspect of the how, Indigenous theorist and writer Leanne Betasamosake Simpson writes in relation to Nishnaabeg thought and practice, "It became clear to me that *how* we live, *how* we organize, *how* we engage in the world—the process—not only frames the outcome, it is the transformation. *How* molds and then gives birth to the present. The *how* changes us. *How* is the theoretical intervention."[125] *How* is the world-making process that traverses ways of life, modes of politics, registers of experience, and integrities of relation. It forms subjects and environments in its indication toward engagements. It is both a theoretical orientation and an embodied collective practice.

Expanding Practices of the How-To

When using sensors in the open air, the question of how to assemble toolkits expands into other orders of instrumentality and the how-to. Is this just a matter of distributing air-pollution sensors to a community? Or is this process a way of forming new networks and communities of inquiry? And if it is the latter, then how might it be possible to expand upon the makerly way of encountering sensors to engage with these devices as more fully social technologies that are constituted in and through diverse more-than-makerly social environments? How-to is a question that activates and indicates how to address that problem.[126] The how-to of citizen-sensing toolkits frames the problem of environmental pollution as one of measurable quantities that can be documented and communicated as evidence. Yet this how-to also organizes an expanded set of practices, from how to build a community-monitoring network that responds to the sited problem of pollution to how to draw on community expertise and connections and how to gather observations and experiences of environments over time.

As one example of an approach to exploring this expanded configuration of the how-to, Citizen Sense built upon its ongoing practices of meeting with community groups and residents concerned about air quality by developing a "Logbook of Monitoring Practices." The logbook, discussed further in chapter 2, sought to approach the how-to through collaborative research and action.[127] The logbook was one part of the Citizen Sense Toolkit that organized techniques to constructively and collectively ask about the how: how to build a toolkit, how to use sensors, how to monitor, how to use data, and how to effect improvements to environments and environmental pollution. This was not a process that started from a preformed assumption about what technology is or ought to be; rather, it asked what it could become within a community of inquiry committed to collective engagement with environmental problems. You might find that by asking questions about the how-to with a more low-tech device, such as a logbook, it can be possible to configure an expanded toolkit through a process of collective research.

In this way, our first "Logbook of Monitoring Practices" was developed as a series of questions to ask participants about how they would document the problem of air pollution from fracking. These questions were entry points into the how-to: they asked how to monitor this complex and fraught environmental problem that people had struggled with for years. These questions could be worth considering when developing your own toolkit that seeks to build community-monitoring networks. The questions include the following:

1. What pollutants should be monitored?

From benzene and carbon monoxide to particulate matter, nitrogen oxides, ozone, light, and noise, many possible pollutants are associated with the industrial process of fracking. This set of questions asks what the primary pollutants of concern are and what the toolkit should include to document these pollutants.

2. Where should monitoring take place?

Pollution can occur throughout the hydraulic fracturing process and across its infrastructure, including at drill sites, well pads, compressor stations, glycol dehydrators, impoundment ponds, and pipelines. Here the logbook maps where the most noticeable emission sources are. It also queries what other pollution sources might not be monitored or regulated.

3. Who is monitoring?

Some monitoring activities might already exist and could be undertaken by government agencies or industry. This set of questions asks who might be monitoring

Figure 1.10. The first version of the Citizen Sense logbook used to gather contributions about existing and proposed environmental monitoring of fracking and air quality in northeastern Pennsylvania. Photographs by Citizen Sense.

already, who should be monitoring if this is not taking place, and how the data should be made accessible.

4. What monitoring practices are citizens already undertaking?

When pollution is suspected to be occurring, it is common for residents to begin monitoring to determine whether their air, water, soil, and surrounding environment could be contaminated and causing harm. By learning more about existing monitoring practices, it is possible to incorporate these knowledges and experiences within the development of expanded toolkits.

5. What exposures have been noticed or felt?

By inquiring about how exposures are experienced and the distance between natural gas infrastructure and homes, it can be possible to understand the health effects that could be linked to emissions.

6. What is difficult to monitor or cannot be recorded?

The fracking process involves many undisclosed substances in drilling fluids, surfactants, slurry, lubricants, and foaming agents. This question asks about uncertainty regarding environmental pollution as well as the possible limits of monitoring equipment for detecting different substances.

7. How should citizen data be used?

Sensors can generate considerable amounts of new data. Large data sets can accumulate when this is multiplied across a community-monitoring network. This question probes how these data could be used, what effect they might have, and whether and how the data should be shared across the community or farther afield.

8. What does a day in the life with fracking look like?

By asking participants to document what everyday life with pollution involves, it can be possible to record the many activities that could be causing pollution and its associated effects. Everyday life might also have shifted in response to ongoing industry operations, and this question searches for observed changes to environments over time.

9. What monitoring scenarios should be tested?

When monitoring different components of fracking infrastructure, distinct monitoring setups could be useful to investigate. For instance, it could be worthwhile to monitor the "life of a well" as it is graded, spudded, drilled, and finished as a well pad producing gas. Monitoring might also take place at set distances from the emissions source. This open-ended question asks participants to consider what a monitoring experiment could look like, to develop a research design, and to put it into action.

10. What additional observations can be made?

Because residents observe changes to environments over time, and witness the effects of pollution as they take hold, you might find it is useful to ask for photographic documentation of environmental changes and any additional observations or questions that can inform the how-to of the toolkit.

These logbook questions for composing a monitoring toolkit are less an absolute list to follow and replicate and more a provisional map of how different questions—questions that ask how to rather than instruct how to—can assemble a process of inquiry, a monitoring toolkit, a set of environmental observations, an indication of how to work with data and evidence, and an understanding of community networks and interests.

Participants' contributions included lists of pollutants to monitor, such as particulate matter, volatile organic compounds, methane, nitrogen oxides, and noise. These were important starting points for how we then came to build toolkits to be installed near infrastructure. In the logbooks, clear indications emerged of what parts of the fracking infrastructure were of particular concern, including compressor stations and well pads. In addition, new information surfaced that might have been overlooked about problems with traffic, including industry trucks, heavy equipment, and helicopters, all serving as the moving infrastructure for hauling fracking materials and waste to and from sites. Logbook contributions offered detailed suggestions for who should monitor. They also proposed different monitoring scenarios, including an installation that would encircle industry sites with monitors and provide real-time data to the community.

The range of environmental events, changes, and pollution that participants added to the logbooks became a complex if informative record for considering how to make a relevant citizen-sensing toolkit. As one participant documenting "a day in the life with fracking" noted, the experience of fracking was characterized by:

Trucks, trucks, and more trucks.
Traffic tie-ups *much* more frequent.
Dust blowing everywhere.
Hills crisscrossed with pipelines.

Slow super heavy equipment on oversized trucks.

60' wide swaths of trees coming down to make roads.

Country lanes being widened and built up, completely changing the character.

Bright lights in the sky at night, near and far.

More helicopter traffic.

Noise from drilling, trucks, *flaring* and compressor stations at all hours.

More litter on the roadways.

Torn up roads.

Torn down barns.

Not safe to ride a bike on the back roads anymore due to trucks barreling around curves.

No more rhythm to life—no downtime. No weekends or holidays. Industrial intrusion 24-7-365.

Neighbors uncomfortable at best, fighting with each other at worst.

New hospital, donations of gas money to all kinds of causes.

People spending money on trucks, tractor, pools, additions to their homes.

Downtown stores going out of business.

Huge staging sites with parking lots full of trucks, equipment, temporary buildings.

This iterative and collaborative process of asking how-to and of gathering collective observations informed the development of the Citizen Sense Toolkit. The toolkit formed as a collection of different air-quality monitors that participants used to gauge air quality from hydraulic fracturing activities. Yet it was clear that there was much more to undertaking a citizen-sensing project than distributing monitoring technology within a community. The logbook became a toolkit within a toolkit for learning more about the existing networks of monitoring and action as well as the sedimentations of pollution, politics, and conflict within a distinct area. It framed the how-to as a series of questions, which in turn attended to the communities of inquiry that had formed, and could be in the process of forming, through a citizen-sensing project to study air quality.

From Makerverse to Pluriverse

Making is often discussed as a good or end in and of itself, especially in the sphere of digital technology. Action, getting practical, building, working in a hands-on way: these are proposed as remedies to more sclerotic and inert—indeed, even "academic"—approaches to problems. As Lily Irani has observed about an account of a hackathon, there was a notable "bias for action" in the planning for this event, where hackathon participants "sought to intervene in the

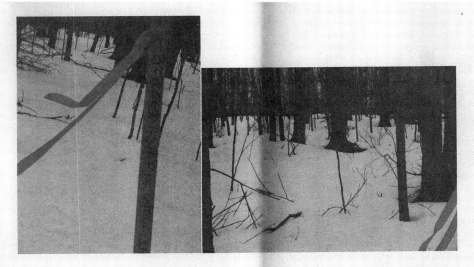

These pictures were taken at the top of the hill behind our house two properties down. I used to walk and snowshoe up along the ridge several times a week before the drilling started. The neighbor in between us and this property also has a non-surface access lease as we do. This was the first sign of the drilling presence at the Knapik sight. This picture was taken March 5, 2011.

This is a stitched together panorama of the same woods from pictures taken November 28, 2011.

Figure 1.11. Participant contribution to logbook documenting before-and-after tagging and clearing of wooded area to construct a fracking well pad. Photographs by anonymous Citizen Sense participant; courtesy of Citizen Sense.

operations of the world through 'action' and 'making.'"[128] Such emphasis on making and action can constitute a makerly subject—or instrumental citizen—who
undertakes activities because they seem productive. Yet, as the pragmatists have
discussed and critiqued, action for action's sake is an empty project. Practice is,
notwithstanding, the space within which ideas are put to work. It is the very
operationalization of ideas—the instrumental aspect of instruments—that the
pragmatists stressed was key to how inquiry unfolded and came to have effects,
less as the proof of a theory and more as contingent and concrete experience.

In a different but parallel register, Simpson notes that distinct approaches to
making are part of the integral connection between Indigenous knowledge and
practices. Making can be "the material basis for experiencing and influencing the
world."[129] Her discussion of making and theory is tied to Indigenous contexts,
and it also produces a philosophy that differently resonates with the pragmatists'
approach to practice. Yet Simpson further draws out how the collective undertakings of Indigenous politics and governance are also embodied and implemented
through distinct forms of making, politics, and governance. She writes:

> Kinetics, the act of doing, isn't just praxis; it also generates and animates theory
> within Indigenous contexts, and it is the crucial intellectual mode for generating
> knowledge. Theory and praxis, story and practice, are interdependent, cogenerators
> of knowledge. Practices are politics. Processes are governance. Doing produces
> more than knowledge.[130]

The more-than-knowledge that doing produces involves the very relations and
networks that make worlds—and these are political inquiries and inhabitations.
Just as making does not take place simply for the sake of making or action, doing
is about more than a refinement of theories. Doing unfolds ways of being in and
being for worlds. Doing can reproduce practices such as settler colonialism. But
it can also test, transform, and generate theories in a connected pursuit of the
how-to that works toward more breathable worlds. As forms of doing and action,
instruments and instrumentalities are not, in this way, direct lines to certain outcomes but rather constitutive and contingent operations that form worlds.

The imperative mood resurfaces here, less as instruction and more as procedure and practices that form networks and worlds. Procedure is always open
to revision through ways of living in and making worlds. Making involves differential ways of being in worlds. Making, action, practice, and procedure generate worlds in the plural, the pluralistic universe—or pluriverse—that was the
focus of James in his work on radical empiricism.[131] Here, instrumental citizens

form with and through practices that would transform polluted environments by working toward more breathable worlds as a contingent and collective project.

As the "Logbook of Monitoring Practices" example demonstrates, there are multiple ways of monitoring environments and accounting for the effects of pollution through forming toolkits, making sensors, identifying monitoring scenarios, and gathering and analyzing data. How-to can be a way to recognize and support a plurality of modes of inquiry, technical practices, and environmental relations. The makerverse of DIY technologies shifts to become the pluriverse of reworked toolkits and action. The instruments and instrumentalities of sensors are not a unidirectional unfolding of makerly agency but rather networks-in-formation that generate collective effects.[132] Citizen sensing unfolds not just through sensor devices but also in concrete locations and as collective monitoring projects for documenting and addressing environmental pollution. When building a network, you might find it helpful to remember that this is an ongoing practice involved in pluralistic fields of relations.

HOW TO TEST RESISTANCE

Once, while presenting Citizen Sense research, I was asked by an event participant whether the work was somewhat risky to undertake because it could be perceived as an "activist" project. Indeed, the questioner considered the topic of fracking controversial and suggested that "helping people" would forego the objectivity that is meant to characterize academic research. I have received variants of this question in several other contexts. The gist of these inquiries is a worry over the loss of expertise that is seen to be granted by being a distant academic observer and commentator, ideally working on a more neutral research topic.

If ever there were an anecdote well aligned with feminist technoscience, this one surely must seem ready-made to demonstrate the relevance of this body of work. Cue Haraway's "modest witness": the very perception that inquiry involves standing back and letting events take their course, whether in the form of instruments and air pumps or social and political affairs, is a gendered and privileged way of organizing inquiry that allows some people and actions to recede from view to generate universality and objectivity, while others are branded as illegitimate because their presence jams the signal of objectivity.[133] It would be similarly possible to pass through the quantum feminism of Barad to articulate that any observation—even the seemingly most technical and scientific—is an achievement that involves sociopolitical relations.[134] And traveling back to the formation of quantum theory, along with its influence on theorists, including Dewey, it would also be possible to say that observing and acting are involved with each other. Observing is acting. Rather than assuming the position of nonengagement to achieve objectivity, Dewey (under the influence of Heisenberg) suggested that new modes of engagement should be deliberately sought to pursue the promise of instrumentalism and philosophy as action.[135] Indeed, as West has pointed out, for Dewey this was a way to ensure "active engagement with the events and affairs of the world" that would contribute to "a worldly philosophy and a more philosophical world."[136]

You might find that, when taking sensors into the open air, working to build community-monitoring networks, and grappling with environmental problems, resistance takes on an electropolitical oscillation. As communities diversely engage in environmental struggles, sensors enter into the fray as part of a process of inquiry and evidence making. Indeed, struggle is central to how these projects and practices unfold. Resistance will be encountered not just as a lesson in voltage but also as a response to citizen data, as a query about proper modes of research, and as questions about how or whether governments could be more accountable. Resistance will also be cultivated through processes of circumventing established ways of dealing with or overlooking environmental problems, gathering and presenting evidence consistently and insistently when it is ignored, and organizing meetings and listening sessions to make citizen observations of environmental problems matter. At the same time, it is important to account for how expertise differently manifests and how this informs citizen-sensing technologies and practices in the attempt to struggle with accounts of environmental pollution.

The label of activism suggests that the research has forgone its potential for legitimate inquiry. Yet, on the contrary, because the research is working with and through action and engagement, it is developing new capacities and open-air instrumentalities. Despite the marketing promises, sensors do not simply deliver transformed political engagements or environmental solutions. Sensors neither singularly empower people nor instantly transform them into activists. Instead, research and practice that are variously situated as collaborative or participatory demonstrate that engagements with environmental problems unfold through differential and complex political struggles. By undertaking practice-based and collaborative research, such "findings" become evident. This can also be a way to begin decolonizing research practices and rework the expert–citizen relations that colonial modes of research can fix into place.[137]

When working with communities in a participatory way, it is possible to learn about the multiple approaches to addressing environmental pollution, the friction and the discord, the diverse strategies for organizing, and the environmental encroachments that have been held at bay. It is also possible to better understand how collective politics materialize less as a singular pursuit of a goal and more as a working and reworking of instrumentalities: there is work to be done, but the doing of it causes new actions and relations to form. In this way, the political subjects—the instrumental and active citizens—that are constituted through these modes of action are in process. Drawing on the previous discussion of pragmatism, it would be possible to say that action is not the elision of conceptual reflection or development but rather its test and fulfillment. While in the pragmatists' estimation, action is not to be pursued for action's sake, it is also possible to ask what modes of action are underway and what experiences and worlds these would generate.

Activism is one way of parsing action in politics, and yet there is often disagreement about what does and does not count as activism. The how-to is also about modes of action and calls to action. This action can be parsed in many different ways: as action for action's sake or as materialized ways of living. Modes of action are also shifting in response to present demands. Rather than that old Leninist question "What is to be done?,"[138] a more usual question now starts with "How to . . . ?" For some, the question is simply a version of "How to make . . . ?" For others, the question is "How best to live on, considering?"[139] This latter question, raised by Berlant, is an appropriate one to dwell on at this juncture in this text because this modality signals most clearly the struggle and resistance that can be embedded within or activated by the how-to.

The search for instructions, the following of procedures, the hopeful pursuit of an effective action or promised outcome: these are ways of looking for direction when potentially floundering on the shores of life. You might find that how-to is a mode of action that often starts in the imperative mood and follows an instrumental trajectory. But how-to is also a vector of transformation. It generates open-air instrumentalities and other ways of undertaking research as a collective project. Transformation, nevertheless, encounters resistance and requires struggle. Testing resistance, then, is an important way to keep your toolkits well tuned and ready for diverse modes of action, and even activism.

HOW TO RETOOL ACTION

I am frequently asked whether Citizen Sense research is empowering people and communities through the participatory research undertaken using citizen-sensing technologies. The short answer is, not as directly as that. The medium answer is, it is best to query the uncomplicated connection between technology and empowerment. The long answer is, it might be advisable to review this how-to

Figure 1.12. Example of a logbook-based citizen contribution documenting an EPA mobile monitoring unit that undertakes periodic environmental monitoring. Participant contributions were returned by email and via SD cards mailed along with logbooks back to Citizen Sense. Photograph by anonymous participant; courtesy of Citizen Sense.

exploration, which seeks to trouble and retool how toolkits—and empowerment—constitute trajectories to action. Or, as Isabelle Stengers has suggested, it might also be possible to consider how to undertake the "empowerment of a situation," which involves "giving a situation that gathers the power to force those who are gathered to think and invent."[140]

This chapter as how-to guide has explained how it might be possible to inhabit yet also to transform how these technologies operate through open-air instrumentalisms, and to seek out creative forms of misuse that challenge the assertion that technology made this happen or that Sensor = Outcome. In this way, it might also be possible to engage with and yet reorient the usual and primary attention away from how to make a sensor talk to a platform toward more open-ended engagements with these technologies. Such reorienting and retooling practices challenge the usual configurations of action—and empowerment. They also rework the political relations that are made possible. This is a way to retool toolkits and the action they would organize.

When using the term *retool*, I am inevitably drawing on work from feminist theory and technoscience to propose how to work against the grain of dominant technological narratives.[141] Retooling is a practice honed through struggle: struggle with and against standard operating procedures. Retooling is a way to transform and invent technoscientific practices. It asks how toolkits as trajectories to action are identified, how they are operationalized, which subjects are drawn into their modes of action, which relations are configured, and which worlds are made and sustained. These questions of process and mechanism—the "how" of the how-to—are asked so that further engagement and working through of instruction and procedure might find the flex points for transformation.

While this text could have initially undertaken a survey of citizen-sensing and citizen-science toolkits to introduce this field of study, I have deliberately opted not to follow the categorical impulse but instead sought to examine the imperative mood and instrumental modes of action. Rather than pursue a definitional or taxonomic study of practices or toolkits, this text suggests that toolkits generate open-air instrumentalisms that transform possibilities for political engagement and action. Categories could only ever serve as a provisional way to understand these toolkits and practices.[142] Indeed, the more-interesting toolkits incorporate contingency as a crucial part of how they provide resources for organizing action.

Many other toolkits, DIY projects, and community projects have traversed this space of instruction, contingency, action, and alternative engagement. From the Center for Urban Pedagogy's "Making Policy Public" pamphlet series, which explains and guides publics through complex legal issues like housing and workers rights,[143] to *A Guidebook of Alternative Nows*, which collects examples of alternative economies and engagement experiments,[144] to Zach Blas's manual for "queer technological strategy,"[145] to the *3D Additivist Cookbook* for cooking up alternative inhabitations in troubling times,[146] and the Detroit Community Technology Project guidebooks,[147] a wide range of toolkits and guidebooks are experimenting with the form of the instructional and the imperative to work toward more democratic operating conditions.[148]

There are also very different ways of engaging with DIY and technology that can contribute to community projects of addressing environmental pollution and public health. Alondra Nelson has described how the Black Panther Party undertook projects in DIY community health activism that offered alternative means of mobilizing medical technology and political subjects in the interests of social justice.[149] These practices offer distinct ways of transforming technologies and social relations. They make alternative worlds through attending to the political subjects and communities of inquiry involved in open-air instrumentalisms, where experimenting with the conditions and potential of altered technoscientific arrangements can also undo power structures that contribute to health inequalities.

The point of interrogating instrumentality in this way is to consider how citizen-sensing technologies could be described as instrumental in the limited sense: as merely functional and utensil-like. Politics, as scripted through these engagements, could also be seen to fall into the trap of a more reductive instrumentality. But as this chapter suggests, there is more to an understanding and practice of instruments and instrumentalities than might initially have been suspected. Instrumentalisms become prospective in the effects they generate and the relations they inform. They take shape in the open air, as open-air instrumentalisms. Instrumentalities generate new political inhabitations. The toolkit and the instructional are not necessarily expressive of the starkly functional or extractive form of instrumentality, because instruments develop through engaged and contingent practices. Instrumentalism involves setting in motion, operationalizing, and potentially transforming. Instruments—whether in the form of concepts or sensors—are instrumental to the unfolding, the doing, and the transforming, where other ways of living and other processes are articulated.

I have suggested here that "how to do things with sensors" is a project that moves from the imperative mood and reductive instrumentality to one that might generate more contingent open-air instrumentalisms, particularly in relation to citizen engagement with environmental monitoring technologies. This reworking of citizen-sensing technology and technical relations intends to counter the sinister veneer of Silicon Valley and the smug tyranny of the tech bro, where normativity, exclusion, and reductive technical relations contribute to unjust and undemocratic practices, relations, and worlds.

Once you start to look for instruments, you might find them everywhere: much like Haraway's air pump, they are at work in-forming and re-constituting citizens and worlds. This guidebook suggests that it would be advisable to approach these instruments through the concept and practice of open-air instrumentalisms, where experimental approaches as well as new technical relations, modes of inquiry, forms of political engagement, environmental commitments, and ways of making breathable worlds might materialize.

In this sense, this chapter interrogates the diagram of citizen sensing as a mode of technological engagement that leads to specific political effects. Drawing on pragmatism, feminist technoscience, environmental justice, Black studies, and Indigenous theory, this how-to guide develops an approach to sensors where methods, practice, ideas, and theory are co-constituted, embodied, and retooled. To ask how to do things is to ask how to transform things. It is to inquire how to experience and influence worlds. Instruments and instrumentalities do not offer up guaranteed ends; rather, they unfold operations that are ways of engaging with ideas, technologies, relations, entities, environments, and worlds. Dewey referred to instruments as ideas capable of "organizing future observations and experiences" rather than "reporting and registering past experiences." Instrumentalisms, in this sense, are propositional. If there is action to be undertaken, they are in some way focused on making action and change possible. This is an approach that focuses on "consequent phenomena" rather than historical facts, and is what Dewey would refer to as something "revolutionary in its consequences."[150]

Here's what you might have learned in the process of following these instructions: how-to is a proposition, not a rule. Its imperative mood is one of responsibility and even urgency more than command. How-to is an instrumental project, where meaning arises through contingent operations that make and remake breathable worlds. How-to enables open-air modes of inquiry, action, and conduct. How-to is experimental in its searching after ways to address problems. How-to demonstrates how distinct ideas and instrumental actions are tied to different communities of inquiry and possibilities for transforming and retooling action. While the how-to might initially seem to present straightforward instructions pointing toward guaranteed results, the how-to approach should necessarily engage with the pitfalls, deviations, and anti-triumphalism of undertaking citizen-sensing and environmental monitoring projects in concrete situations. Such a how-to toolkit, then, is productive of open-air instrumentalisms. It works within the genre of the how-to but also seeks to retool this narrative and trajectory toward action in order, as Berlant suggests, "to invent new genres for the kinds of speculative work we call theory."[151]

In the chapters that follow, even more how-to practices proliferate. These instructions and procedures span from how to monitor pollution over time, to how to learn atmospheric chemistry, how to analyze data, how to construct evidence, how to ring a regulator, how to influence policy, how to organize a movement, how to remake environmental relations, and how to make more breathable worlds. Many more how-to inquiries beyond this unfold in the upcoming discussion, where citizen-sensing practices unfold in concrete situations to demonstrate how political subjects and relations materialize as uneven and yet lively formations of citizens and worlds.

TOOLKIT 2

FRACKBOX TOOLKIT

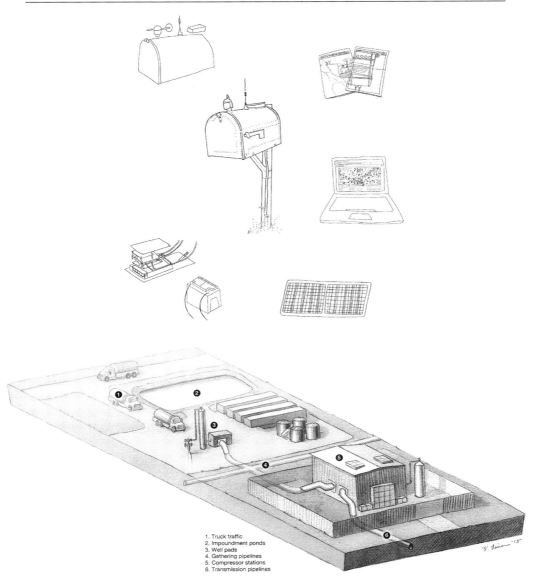

1. Truck traffic
2. Impoundment ponds
3. Well pads
4. Gathering pipelines
5. Compressor stations
6. Transmission pipelines

Frackbox Kit developed by Citizen Sense for monitoring air quality in relation to fracking infrastructure, northeastern Pennsylvania. Illustration above by Sarah Garcin, illustration below by Kelly Finan; courtesy of Citizen Sense. This toolkit can be found in a more extensive form online at https://manifold.umn.edu/projects/citizens-of-worlds/resource-collection/citizens-of-worlds-toolkits/resource/frackbox-kit.

Chapter 2

SPECULATIVE CITIZENS
How to Evidence Harm

In the Endless Mountains of northeastern Pennsylvania, residents and community groups are monitoring the growth and impact of a relatively new industry, hydraulic fracturing. Also known as "unconventional shale gas extraction," or simply "fracking," this industry increasingly crisscrosses and carves up Pennsylvanian landscapes, as well as many other sites around the world, from Oklahoma to Siberia. Fracking involves extracting natural gas through first drilling vertically thousands of feet underground, then drilling laterally up to a mile and a half beneath shale rock formations, and finally injecting vast amounts of water, sand, and chemicals to fracture shale deposits and release bubbles of gas trapped in the porous rock. The extensive infrastructures of fracking span well development and drilling, well completion and production, on-site and off-site processing, distribution and storage of gas.

At every point in this infrastructure, pollution potentially occurs to air, water, and soil. The wells drilled at initial points of extraction generate greenhouse gases primarily in the form of methane, along with air pollutants including particulate matter and volatile organic compounds (VOCs). The water and chemicals used to exert pressure to remove shale gas can contaminate drinking water and surface water through wastewater ponds of "flowback" drilling fluid left to be trucked away or evaporate into the air and settle into the soil. The compressor sites where gas is pressurized, refined, and pumped into pipelines generate additional methane, diesel, and VOC emissions in the form of benzene, toluene, ethylbenzene, and xylene (BTEX compounds), some of which are known carcinogens at even minute levels of exposure.[1] And the extensive truck traffic that hauls materials for initial well development to waste removal contributes to ultrafine particulate matter and diesel emissions recognized as carcinogens by the World Health Organization.[2]

Figure 2.1. Fracking well pad undergoing active drilling; oversized truck transporting fracking equipment in northeastern Pennsylvania. Photographs by Citizen Sense.

Across these infrastructures of energy extraction, established, new and un-certain formations of pollution occur that are yet to be studied both for their distribution and type and for their possible future accumulations and effects. The rise of fracking has forced multiple questions about how this mode of energy production could damage environments and health. In the United States, shale gas has been referred to as a "bridge" technology. Widely considered cleaner than coal, it has been promoted as an interim solution on the way to more renewable and sustainable energy sources while also reducing dependence on imported energy sources. Yet the extraction, distribution, and use of these fossil fuels cause concern about how to evidence bodily stress and environmental pollution from these industries.

In Pennsylvania, residents take up assorted instruments to monitor and doc-ument fracking-related pollutants in the air, water, and soil. People use a battery of equipment, from badges to sensors and video, to detect and track pollution. Some of these devices are analog and low-tech, while others are digital and more complex to operate. They also share their techniques and findings with other community groups, state and federal regulators, and environmental and health NGOs. By using monitoring devices, residents seek to evidence harm. They express care for environments and health by documenting pollutants in situ and attempting to link these ill effects to health and environments. In a different but resonant register to the NoDAPL drone video monitoring in the Introduction to this book, fossil fuels that crisscross landscapes produce ongoing pollution and contamination, which further generates an array of citizen-sensing practices that track pollution to hold polluters to account.

One such digital technique involves using forward-looking infrared (FLIR) video documentation. When the FLIR camera operates in non-infrared mode, emissions from gas infrastructure are not visible. However, when in full opera-tion the FLIR thermographic camera exposes significant drifts of methane, VOCs, benzene, and other gases that are often leaking from gas infrastructure.[3] In black-and-white or magenta-and-orange scenes framed by the FLIR menu and time stamp, leaking emissions that are otherwise not detectable become evident. FLIR thermal imaging is typically used for safety checks and monitoring hot spots. Here, citizens put this technology to work to observe the emissions from a com-pressor station as one point in the more extensive sprawling infrastructure of the natural gas industry.

People post FLIR videos on YouTube that document emissions from fracking infrastructure in Pennsylvania and farther afield. In a video recorded by Frank Finan at a compressor station in Dimock near the Tennessee Pipeline, vertical and L-shaped exhaust pipes discharge emissions at high velocity.[4] Slower vaporous

Figure 2.2. Frank Finan monitoring emissions at a compressor station using a FLIR camera. Photographs by Citizen Sense.

clouds surround more rapid jets of emissions, with some plumes billowing and trailing wistfully and others carving stark vertical lines. The camera pans and fixes on numerous outlets, although it is often difficult to determine the precise location of leaking emissions. The image is framed with the markers of the FLIR camera details. The time stamp, 5/21/14 at 3.35.45 pm, advances for 2 minutes and 2 seconds of footage. The battery indicates that it is partially charged. The FLIR menu functions mark the top edge of the frame: HI OFF, AUTO, HIST, BL. Soundless and silhouetted infrastructure pumps and heaves fumes skyward. Gas deep underground hurtles and shuttles upward to wellheads, then on to pipelines, which channel gas to this point to compress and equalize it to be sent farther along pipelines and to additional compressor stations. Impurities are removed, and explosions are meant to be prevented, so periodic "blowdowns" are staged to release excess pressure from the compressor site.[5]

Frank began using the FLIR camera to document the undocumented and often poorly understood effects of fracking. With footage dating back to 2011, his videos, along with records from many other residents in the area, form an archive of citizen environmental monitoring.[6] They track the changing landscapes of the Endless Mountains and many other shale plays across the United States. At the time of this writing, Frank had uploaded nearly one hundred videos that document emissions at compressor stations and pipelines, in wooded settings and farmlands. The FLIR documents this infrastructure as it undertakes the work of fossil-to-fuel conversion, where once-ancient plant matter surfaces to the atmosphere in particles and gaseous compounds: methane, VOCs, carbon dioxide, nitrogen oxides, particulate matter, and more. The citizen videos record emissions as they surge and leak from vents and release valves, altering atmospheres, affecting bodies, and transforming environments.

As *Citizens of Worlds* documents, air pollution is a problem that causes significant harm to health and environments. The WHO has established that air pollution is one of the leading causes of disease and death worldwide.[7] Air pollution also connects to environmental struggles. It is an indicator of resource extraction, rampant development, fossil-fuel consumption, traffic gridlock, and contentious infrastructure projects. Air pollution can be evident at distinct sites of spatial segregation, as environmental justice research has demonstrated,[8] and it can also be much more pervasive within congested urban environments. As will be discussed in the next chapter, nearly all Londoners experience harmful levels of particulate matter. Air pollution can be omnipresent and yet unevenly distributed in the harm it causes. Breathing and "breathtaking spaces," as described by Christina Sharpe,[9] are particular practices and sites worthy of study to understand how people experience environmental and social injustices, how they cultivate practices of combat breathing, and how they work toward more breathable worlds.

Air-quality monitoring ordinarily takes place through distributed infrastructures meant to protect public health by lessening the effects of air pollution. From health research to policy guidelines to official monitoring stations and labs that analyze data, these monitoring infrastructures can inform corrective action, typically through policy measures, if levels of pollutants exceed guidelines. Air-pollution monitoring could be approached as a material expression of governmental care for public and environmental health. Yet care can as likely turn to neglect and harm, since instantiations of care may be incomplete and even lead to forms of oversight and inertia, where worlds become more or less breathable for some and not others.

In rural environments where most fracking occurs, there is a relative absence of air-quality monitoring networks, because air pollution is generally seen as a problem of urban environments and higher population densities. At the same time, in the United States, fracking is relatively exempt from federal-level clean air and water regulations (in the "Halliburton loophole" of the US Energy Policy Act of 2005). As an industry, it is not subject to the same national safeguards that might prevent pollution to air and water, since these regulations are mainly devolved to states.[10] In this sense, there are many ways in which exposure to harm might not be monitored or prevented, whether through lack of policy or regulation of pollutants and industrial processes or because individuals experience distinct and situated exposures to pollution that the typically fixed and sporadic monitoring stations cannot document. Government-run environmental monitoring infrastructures then materialize as uneven distributions and enactments of care.

Multiple citizen-based and scientific monitoring practices have taken place in the Marcellus Shale region to address the relative lack of data about air pollution from fracking. These practices create alternative monitoring infrastructures to document harm and address the relative lack of care for environments and health. This chapter investigates how residents, activists, and community groups deployed multiple monitoring technologies to document fracking-related environmental pollution and address gaps in regulatory approaches to pollution. In the course of this chapter, I consider how citizen-sensing practices support different ways of evidencing harm and materializing care as expressions of citizenship. Environmental monitoring with environmental sensors could both facilitate and limit this process. In this way, practices for sensing pollution give rise to speculative citizenships and struggles that work toward more breathable worlds.

While a certain amount of attention has been directed toward citizens' monitoring of water quality contaminated by fracking because of the spectacular and

Figure 2.3. Participant contributions to logbook documenting fracking infrastructure and pollution, including flaring at a wellhead, and construction activity at a drilling site. Photographs by anonymous Citizen Sense participants; courtesy of Citizen Sense.

alarming phenomenon of residents in fracking sites being able to light their water taps on fire due to high levels of methane migrating from potentially faulty well casings,[11] this research focuses on the relatively under-examined topic of citizen sensing of air pollution at fracking sites.[12] Whether in the form of ultra-fine particles and particulate matter, nitrogen oxides, climate-change-accelerating methane, VOCs, ozone, and more, an array of compounds generated and following on from fracking processes are known to be accumulating in the air and suspected of affecting bodies and environments.

Following on from the last chapter, which investigated how to retool action through the instrumental citizen, this chapter looks more closely at the speculative citizen. I address practices of citizen-based monitoring of air pollution near fracking sites as speculative attempts to evidence and address harm to environments and health. Citizen sensing of environmental pollution can unfold through speculative registers, because it seeks to generate alternative or supplementary forms of evidence while transforming political engagements for addressing environmental harm. Speculative forms of citizenship potentially materialize through environmental sensing practices as they search for prospective forms of political assembly, engagement, and effectiveness that have yet to be realized.

This chapter attends to how residents' experiences and anticipations of harm have contributed to practices of monitoring environments. The Citizen Sense research group documented and reviewed these monitoring practices, while at the same time working with residents to develop and install citizen-sensing toolkits throughout a three-county area in northeastern Pennsylvania. In this context, I investigate toolkits as they are developed, installed, put to work, broken, and queried in the open air. This practice-based and collaborative approach expands the discussion of the how-to and open-air instrumentalisms from the previous chapter to consider what unexpected uses, committed engagements, and heated struggles unfold through the use of citizen-sensing technologies.

Multiple modes of how-to engagement surface here, including how to establish the "facts" of pollution, how to sense pollution, how to activate collective practices of open-air inquiry, how to transform data into evidence, how to evidence harm, and by extension, how to mobilize speculative monitoring practices toward transformative political engagements.[13] This chapter next considers how speculative practices for evidencing harm and expressing care generate practices of speculative citizenship. I then describe in more detail how people have monitored environments to document air pollution, and the collaborations we undertook to develop a citizen-sensing infrastructure for tracking industry emissions. "How-to" unfolds here as a process of collective anticipation and inquiry. Such inquiry attempts to document harm and generate speculative practices and infrastructures of care that contribute to more breathable worlds.

SPECULATIVE PRACTICES FOR CARING ABOUT AIR

Citizen-sensing practices for monitoring air pollution are often described as a way to "care about your air."[14] Such practices can seem to offer a straightforward strategy for protecting one's health by avoiding exposure to air pollution. Yet in the absence or inaction of governmental air pollution infrastructures, such practices do not readily generate direct solutions to the problem of air pollution, since they do not reduce overall levels of air-pollution emissions. Instead, caring about air becomes entangled with speculative practices for evidencing and addressing harm. Neither care nor the subjects and actions that would constitute care are so clearly identified, since the forms and forums needed for citizen data to have an effect are in the making, and forms of harm are accumulating and often not fully known. Moreover, the conditions in which these monitoring practices could gain a foothold and demonstrate environmental and bodily harm as experienced and yet to come are in process, forming in relation to lived conditions.

Citizen-sensing practices for monitoring air pollution are ways of expressing care about breathable worlds by attending to exchanges across environments, entities, communities, and health. Practices of collecting air-pollution data are speculative attempts to document harm and demonstrate the need for care. Operating outside of the more official infrastructures of care, citizen-sensing practices indicate that more attention should be given to air pollution. They attempt to instigate corrective actions. Yet the exact contours of these political engagements can be somewhat open-ended, and they do not immediately translate into regulation, policy, or even agreement about common environmental problems.

Proposals and practices of care are not straightforward. Moreover, citizen monitoring of air pollution could be generative of what Lauren Berlant calls "cruel optimism,"[15] a concept that addresses how political hopes can generate self-defeating or threatening conditions rather than the liberation they would promise. Here, technologies seem to generate the care lacking in governmental practices and infrastructures. Yet these same devices could as easily produce overlooked data, failed inquiries, and half-hearted engagements. With these cautionary tales in mind, I shift the focus from making normative proposals for care to addressing the complex and speculative practices of evidencing harm as contingent precursors or entreaties to care in the making. Such an approach is more propositional. It resonates with Puig de la Bellacasa's suggestion that "engaging with care requires a speculative commitment to neglected things."[16] Speculative commitments could refer to practices for expanding potential within present political engagements as well as practices for generating citizens and worlds where other ways of addressing air pollution become possible.

While monitoring ostensibly focuses on gathering the "facts" of pollution, a speculative approach to monitoring involves the co-constitution of facts and worlds where those facts make sense.[17] Rather than accumulating facts as self-evident demonstrations of environmental pollution, a more speculative approach to citizen sensing shifts the conditions in which observation, evidence, and care might materialize. Speculative propositions do not articulate in advance the conditions in which they will have relevance; instead, they bring into existence movements of thought and thinkers, citizen and data, where different inhabitations could be possible.[18] In this way, citizen sensing is a proposition for how to document experience, generate facts, and build worlds in which those experiences and facts are relevant. Here are practices whereby speculative citizens might constitute breathable worlds as exchanges and evidence in the making.

But these speculative capacities extend to more-than-humans; they are not merely an attribute of human citizens as usually understood.[19] Experiences are distributed, and speculation is a practice undertaken collectively. Speculation can be distributed through things, which are propositions and potentialities for feelings and encounters: they lure entities into ways of being. In this sense, any account of "the social" would necessarily need to attend to the multiple entities that are continually sparking speculative encounters. When monitoring for pollution at fracking sites, these entities include sensors and chemical compounds, data platforms and wellheads, truck traffic and meeting halls, bodily afflictions and noxious smells, as well as ancient rock and energy markets. Speculative practices for evidencing harm unfold with and through environmental sensors and these extended milieus. This is where speculation meets open-air instrumentalisms, where the practices of evidencing harm are not simple actions leading to outcomes but involve distributed practices of making citizens and worlds. What might begin as a seemingly straightforward sensor toolkit opens into distributed practices for making more breathable worlds.

Speculation can occur in yet another register, since rather than simply resolve or clearly evidence the probability and effects of pollution, monitoring practices can at times also amplify uncertainty, give rise to speculation, and cause people to wonder, if not worry, about ongoing exposure to pollutants. Some of this uncertainty can proliferate through increased collection of evidence, where the documentation of pollutants can give rise to concerns and questions about effects of pollutants over time, how they will travel through environments and bodies, and whether individuals will find themselves with health issues linked to fracking-related pollutants. Pollution monitoring can activate speculative practices for evidencing harm. These practices could be a way to direct attention to exposed communities. At the same time, speculative practices for evidencing harm could provoke conjecture about future environmental effects, as possible forms of harm-in-waiting that are difficult if not impossible to substantiate. Speculation, here, could be a cause for concern and even dread.

However, speculation neither signals a sort of "relativism" nor forms the basis for dismissing citizen data as speculative conjecture. Instead, it designates how propositions for making worlds come to matter. To dismiss the experience of citizens living on the gas fields would be to fix the environments, experiences, and concerns of fracking as already settled and addressed through a singular reference point of industry, government, or scientific expertise. Yet as with any technology that unfolds in unpredictable ways, new practices for making sense of and attending to this industrial process can also form new collective worlds. Indeed, speculation could be less about resolving uncertainties and more about constituting environments, worlds, and subjects that can register pluralistic evidence and experiences.

Speculative Citizens

By monitoring environments, citizens develop speculative modes of engagement with their lived environments. The instrumental citizen here shifts to the speculative citizen. The "speculative citizen" is a concept that describes how different ways of experiencing environments and pollution assemble as propositions for how to sense and build more breathable worlds. As it turns out, the instrumental might have been speculative all along. Returning to the instrumental experimentalism discussed by John Dewey and influenced by William James, we could say that environmental sensing for evidencing harm is a way of putting propositions to work in the world, in concrete situations. Speculation is not, in this sense, a fictional condition but rather a testing, shaping, honing, and transforming of conditions through open-ended practices of inquiry. Open-air instrumentalisms are both speculative and practical. Their doing and unfolding contribute to the making of subjects, communities, and worlds, as well as political possibilities. These open-air instrumentalisms are not the exclusive work of human makers but are part of a field of influence. This is how worlds and propositions for breathability can take hold.

Speculative citizens are not just articulations of propositional citizenship and political engagement. They are also more ecological and distributed formations of what a citizen is or could become, both as a relational entity and subject informed by exchanges with worlds. Political subjects and worlds materialize through these collective and distributed processes of inquiry. What a citizen or citizenship is or could become is less a matter of definitions and more a question of the propositions, inquiries, practices, and political engagements that might be staged—here through environmental sensing. Yet this operationalization of speculative citizens is not equally available to all who monitor environments or create alternative forms of evidence. Harm might inform what counts as a citizenry worthy of protection or care. Yet it might also exclude some people from making a case for harm, or perpetuate the trials of demonstrating ongoing harm so that people become worn down or exhausted by monitoring efforts.

It is these registers of environmental monitoring that I next discuss through the development and distribution of a citizen-sensing toolkit for monitoring air pollution with residents in Pennsylvania. I document how practices of evidencing harm involve speculative encounters with environments, atmospheres, pollutants, data, regulators, industry, and communities. These practices could be oriented toward attempts to "empower" citizens by shifting the infrastructures, technologies, and monitoring practices to less institutionalized arrangements. At the same time, such practices do not easily or readily mitigate harm. Instead, they require new forms of collective attachment and individuation to activate political engagement and effect. How citizen sensing becomes relevant (or not) then materializes through speculative encounters and commitments to evidencing harm, where the instigation of new relations could be characterized through less normative—and even "complicated"—forms of care.[20] By focusing on the speculative dimensions of evidencing harm, I suggest the processual and collaborative practices of care could be more fully considered.[21] Such an approach focuses on how to generate atmospheric forms of care and environmental policy that are more responsive to multiple experiences and evidence of the harm caused by air pollution, as documented through citizen-sensing practices.

MONITORING FRACKED ENVIRONMENTS

In the fall of 2013, the Citizen Sense project began research on citizen-led monitoring of air pollution on the Marcellus Shale in northeastern Pennsylvania, where there has been a high concentration of active drill sites. The Marcellus Shale is a sedimentary rock formation that spans the Appalachian Mountains and extends across New York, Pennsylvania, West Virginia, and Ohio. The formation is around 350 million years old. An ancient inland sea once settled here, where rock layers and gas from decomposing organic material are now compressed together underground.[22] One of the first shale plays to be drilled in the United

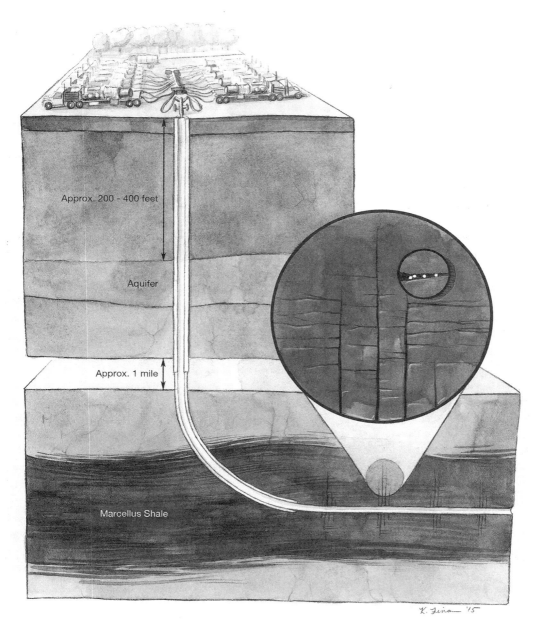

Figure 2.4. Diagram of Marcellus Shale and fracking. Illustration by Kelly Finan; courtesy of Citizen Sense.

States, the Marcellus Shale has undergone its most active stages of development since 2008. As of 2020 in Pennsylvania, this highly productive zone of unconventional gas extraction had nearly 12,450 active unconventional wells in operation, along with 14,666 recorded environmental violations at well sites.[23] Moreover, wells and fracking infrastructure sites continue to multiply. Some estimates suggest the total number of wells will expand to 100,000 over the next several decades in Pennsylvania alone.[24] Environmental violations have included everything from failing to dispose of residual waste correctly, to wastewater discharge, poor construction of pits and tanks, and not adopting Pennsylvania Department of Environmental Protection (DEP) pollution-prevention measures.[25]

Most fracking developments and the leasing of extraction rights are taking place in rural communities with few sources of revenue. Shale gas can boost rural economies by increasing the incomes of retirees and farmers, teachers and local governments. However, at the same time, the rural idyll that may have attracted people to settle here, and the long-standing relationships residents have established with the area, has changed due to shale-gas production and its

Figure 2.5. A Google map created by Meryl Solar to document fracking infrastructure in northeastern Pennsylvania. Photograph by Citizen Sense.

Figure 2.6. Participant showing fracking permissions and infrastructure in Bradford County, northeastern Pennsylvania. Photograph by Citizen Sense.

wide-ranging impacts. This is not to say that this is a pristine landscape, since Pennsylvania is well known for industry such as steel mills, and earlier forms of mining and energy production, including coal extraction. But precisely because there is a prior history of extraction, new extraction economies and practices have brought concerns about what it means to commit now and in the future to these natural-resource and energy economies.

Although attention to fracking's environmental impacts often focuses on the well pads where gas is extracted, the landscape of fracking is not limited to one site. Instead, it consists of an extended infrastructure: horizontal, underground, and emerging at discrete points, interconnected by trucks hauling equipment and waste material, and contributing to airborne and waterborne impacts. Fracking technology uses extensive horizontal drilling with a mix of hundreds of proprietary and often untested chemicals. These chemicals—together with water, sand, and lubricants—are injected into wells under high pressure as fracking fluid to blast out gas from shale layers, which in turn can release methane as well as radiation into the air. Fracking fluid can leach into groundwater and contaminate drinking water.[26] The injected chemicals that return aboveground are stored in impoundment ponds or trucked away to wastewater treatment facilities. The mix of wastewater differs from site to site and can include radionuclides, including strontium and barium, from underground radiation.[27] Fracking is provisional not just because it is an emerging energy technology but also because every fracked site has distinct geological and subsurface features, producing widely different environmental impacts. Yet these impacts are rarely monitored.

In some cases, unconventional shale-gas extraction is referred to (by proponents) as a long-standing technology that has been in use for nearly sixty years. However, other energy researchers suggest that the high-intensity ways fracking is now being undertaken are new, even less than ten to fifteen years old, and have unforeseen and under-studied impacts.[28] As with many technological "innovations," fracking is unfolding as an experiment in the world,[29] where earthquakes, untested and proprietary chemicals, groundwater contamination, and air pollution are among the emerging material-political and environmental configurations and inhabitations that are generated through this mode of energy extraction. This emerging technology contributes to environmental effects that both presently and at some future point could impair living conditions for many within the catchments of fracking operations.

Residents who feel the effects of fracking search for ways to register these impacts on environments and health. Environmental monitoring can be one way to document and evidence environmental change and harm. At the same time, monitoring technologies might not necessarily capture those compounds, events, pollutants, and effects that occupy more liminal, indeterminate, or even unknown and future registers of harm. While individual pollutants can be relatively well studied, their accumulation, amplification, and interaction are less understood.[30] How might it be possible to monitor environments and air if pollutants fall outside the designated list of compounds to monitor and regulate, or if governmental infrastructures are not in place to monitor pollutants? How might it also be possible to monitor and evidence fracking's indeterminate effects, particularly if environmental monitoring detects a select number of substances within a categorical present and does not attend to ongoing interactions of accumulations?[31]

Citizen-sensing practices could challenge more "official" monitoring infrastructures by developing speculative approaches to the uncertain effects of pollution. Different ways of evidencing harm and materializing care can surface through these practices by attending to the overlooked, unacknowledged, newly emerging, and future effects of pollution that could be overlooked by established

Figure 2.7. Fracking protest signs and installations at residences in northeastern Pennsylvania. Photographs by Citizen Sense.

expert-driven monitoring practices. As mentioned in the introduction to this chapter, while governmental air-quality networks could be sited at disparate locations and provide a limited picture of an individual's exposure to air pollution, citizen-sensing technologies intend to provide a more immediate and granular record of pollution. Beyond mapping individual exposure, however, citizen-sensing technologies can produce data where there might be an absence of official monitoring technologies. In this way, sensors can provide alternative data sets to address specific community concerns, such as a polluting roadway or industrial site or the possible pollution of a proposed development. This is a different way of mobilizing public engagements with technology, since communities are not engaged in modes of reflexive deliberation with yet-to-be-introduced technologies and wondering about their potential effects.[32] Instead, they struggle to evidence the uncertain and indeterminate impacts of technological operations as they are already unfolding in lived environments.

In research and fieldwork looking at both scientific and citizen monitoring practices, it has been interesting to note the extent to which atmospheric scientists worry about how citizen-sensing technologies could be deployed in ways that generate inaccurate or unhelpful data.[33] Their concern is numerical accuracy and not compromising the data that would support possible actions attempting to enforce air-pollution policy.[34] However, advocates of citizen-sensing technologies have made the case that the absolute numerical accuracy of the data is of less concern when the process of assembling communities of makers or environmentally engaged "citizens" could be facilitated through the development and use of these toolkits. Others have suggested that data can have increased relevance through the sheer quantity of monitoring underway when distributed across multiple citizen-monitoring sites. Data sets could become relevant by detecting changes in data patterns rather than precise numerical readings. In this way, a greater ability to work with "just good enough data" could be developed to enable situated engagements with environmental problems.[35]

Here, citizen-sensing practices do not necessarily constitute a project of collecting data to raise environmental awareness. Instead, they form more speculative undertakings that register overlooked experiences and exchanges. These practices work toward building breathable worlds, worlds where evidence of harm can register and be redressed. A speculative approach to monitoring air pollution could transform ways of engaging with fracking on the Marcellus Shale. In developing this speculative approach to monitoring, processes of evidencing harm could move beyond an evidentiary tracing of pollutants, whether through high-tech or low-cost instrumentation, to engage with how facts or evidence "take hold" to mobilize relations, practices, and forms of relevance.[36] Monitoring practices

Figure 2.8. Impoundment pond for holding fracking flowback fluid. Residential swimming pool adjacent to new fracking infrastructure. Photographs by Citizen Sense.

are not simply a question of what to sense and how to document pollutants. They also direct attention to some impacts and not others, and inform the possible attachments and expressions of care that monitoring can mobilize. Practices for evidencing harm could generate responsive practices of care, which seek to redress or mitigate harm by attending to impacts that do not ordinarily register. It is these processual approaches to care and practices for evidencing harm that I discuss below.

How to Establish the "Facts" of Pollution

It goes without saying that fracking is a contentious issue on many levels. It can divide communities and create diverging understandings and experiences of pollution and harm. Pollution is unevenly distributed. Residents who live downwind rather than upwind of a compressor site will notice acrid odors and noise. People who live in an area with a contaminated water supply will have to source bottled water. And those who live on a road with constant industry traffic will experience diesel pollution, noise, and congestion at all hours. Many residents in these communities, including those who have leased their mineral rights, have sought to document and understand the impact of these extraction techniques on environments and human health.

People living near fracking sites, compressor sites, waste pits, roads, and other infrastructure have collected evidence of numerous environmental disturbances and health effects, from noise and constant light, to smells from emissions, to a range of symptoms that are characteristic of VOC exposure, as well as asthma and other pulmonary diseases, cardiac diseases, and cancer. Residents near compressor sites notice odors and metallic tastes, which some have suggested are linked to the cleaning fluids used to flush compressors, or to the substances emanating from glycol dehydration processes. Across these multiple sites, residents report experiences of chronic and acute nosebleeds, headaches, dizziness, and a range of symptoms that are difficult to tie into a cause-and-effect logic of how fracking may be affecting environments and bodies. Chronic illness can also take decades to manifest. The ongoing and accumulative health and environmental impacts and harms that could be related to fracking do not always translate into immediate data sets or legible evidence.

The inconsistent occurrences of illness, chemical exposure, and evidence as provided through monitoring make this less a space of demonstrable proof and more an uncertain atmosphere of effects. For instance, tests of drinking water in households where residents complain of illness have at times shown an absence of any substances of concern, and in other instances arsenic, benzene, and heavy metals are evident at high levels.[37] Environmental monitoring does not simply reveal the "facts" of pollution but is entangled with complex environmental, chemical, and bodily interactions. While monitoring might indicate care, care is always yet to be realized, since it requires engaging with the speculative aspects of how harm, evidence, and care could yet unfold.

Indeed, even attempts to generate comprehensive lists of harm often indicate how environmental exposures create uncertainty. The Pennsylvania Alliance for Clean Water and Air has established a "List of the Harmed,"[38] which documents residents in locations across Pennsylvania and the wider United States who have experienced harm from fracking. The list records the specific gas facilities near to which residents live, as well as suspected or evidenced exposures and symptoms for humans and animals. It also includes press and online reports, which can include videos and photographs of harm experienced. Reaching over 23,000 records and 192 pages in length, the list documents residents living next to a compressor station who experience "headaches, fatigue, dizziness, nausea, nosebleeds," with one sample "blood test show[ing] exposure to benzene and other chemicals," as well as the death of goats, cows, chickens, cats, and dogs in areas with contaminated water. Also recorded are environmental nuisances such as seismic testing, noise, dust, heavy machinery sounds and emissions, and "bright industrial lights" throughout the night. The list documents how the light of flaring gas wells can trigger post-traumatic stress disorder, including causing flashbacks for people who have served in the Iraqi conflict.

As a form of evidence, this "List of the Harmed" might be considered to fit within multiple forms of citizen reporting often dismissed as "anecdotal" in contrast to more "scientific" methods for gathering evidence and documenting harm. However, not only is "the science incomplete"[39] when it comes to establishing links between fracking and harm, but residents are often uniquely situated to record their lived experiences of exposure to shale-gas production, and so to contribute different forms of citizen data. Care emerges here by indicating the harm experienced by listed individuals and events, which can inform additional ways of addressing the harm and potential harm experienced by communities. What counts as harm, how it is documented, and how this documentation comes to form evidence are questions about the how-to that citizen-sensing practices similarly generate. Practices of how to evidence harm could become as contingent and responsive as the impacts that they would document and address.

How to Sense Pollution

Practices for sensing pollution involve much more than measuring a pollutant. Questions loom about what to monitor, who is monitoring, and how to act upon monitoring results. Concerns surface about what is unmonitored, unaccounted

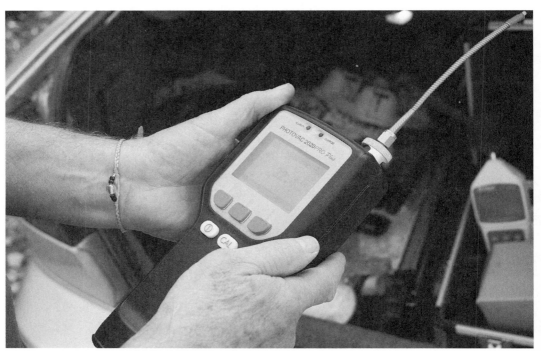

Figure 2.9. Frank Finan showing different devices for measuring VOCs, including a Global Community Monitor bucket for collecting air samples, and a handheld VOC monitor. Photographs by Citizen Sense.

for, yet still could lead to harm. At the same time, the regulatory and enforcement infrastructure for monitoring pollution has not caught up with fracking technology. As a result, it is often ill-equipped to monitor and regulate this industry's complex processes and impacts. Here, speculative forms of citizenship and practices for documenting environmental pollution materialize together as people gather evidence and attempt to make sense of the effects of fracking.

In this context, the Citizen Sense group began fieldwork and desk research in the summer of 2013. We found that many citizen-sensing practices to monitor air and water quality near fracking infrastructure were underway. These practices included various instruments, techniques, sites, pollutants, and environmental media. Residents used devices such as a high-end Photovac 2020PRO Photoionization Detector that can be used for humidity-compensated VOC detection in air, water, and soil. They regularly set up a FLIR Gas Finder that detects seventeen gases at −20°C to +300°C through infrared thermal imaging that, as mentioned in the introduction to this chapter, some have used to document the effects of compressor sites. They participated in installing NGO-loaned summa canisters for testing a range of air pollutants.[40] They set up and wore badges for detecting BTEX chemicals for university studies.[41] And they contributed to bucket-brigade community monitoring, a long-standing analog technique using a bucket with a vacuum-powered pump and bag to draw in air and test for more than seventy VOCs and twenty sulfur compounds—a process that requires samples to be sent off to laboratories for analysis. The data from these citizen-monitoring efforts were collected and presented as lab reports and community organization documents. Many image- and video-based forms of monitoring were circulated online and through video-hosting platforms such as YouTube and Vimeo. Initiatives such as FracTracker provided community-mapping data, and MarcellusGas .org provided monthly reports on fracking production.[42]

While individuals in northeastern Pennsylvania had been undertaking environmental monitoring by using low-tech and high-tech instruments, one of the primary groups contributing to and mobilizing evidence about air pollution was Breathe Easy Susquehanna County. Made up of around twenty members, this citizen group came together in early February 2013 to protect local communities from poor air quality and its health effects, primarily due to the growing fracking industry. In addition to its mission to address air quality, the group outlines its strategy as one of "respectful dialogue between the natural gas industry and our Susquehanna County community." The group's intention was to work with industry to improve air quality across all aspects of the fracking infrastructure, but to encourage voluntary industry efforts in this area rather than seek new legislation. As the chair of the group, Rebecca Roter, writes on the Breathe Easy Facebook page:

> Breathe Easy Susquehanna County PA (BESC) is a fledgling community group attempting to bring together divergent voices who have been pitted against each other around one common concern, air quality. The marcellus [sic] train left the station six years ago in our county. We all live with the same impacts to our community whether we signed a lease or not, whether we were on that train or not. Many of us from across the table share the same concern about keeping our air as clean to breathe as we can. We cannot choose to not breathe as we see more compressor stations and well pads permitted weekly. We need to act now together, to bring our community together now over air quality, to try to keep our air as clean as we can.[43]

The call to work toward breathable worlds, expressed in the very name of this community group—to "breathe easy"—became the basis for developing actions to care for the air, but in ways that would require collective forms of inquiry to establish where pollution was occurring and how to address it.

Air quality was an increasing focal point both for this community group as well as for multiple other residents in the area who were engaged in diverse projects and initiatives to address air pollution. Some residents felt that Breathe Easy's attempt to work with industry but not advance regulation did not hold fracking operators to account. Other groups and residents focused on development plans and used environmental data to contest further industrial activities. While not the only group concentrating on air quality, Breathe Easy was especially vocal about this issue and worked with NGOs, including Shale Test and Earthworks, to collect environmental data on pollution from fracking. The group's members had contributed to VOC testing with buckets and badges and had purchased their own array of monitoring instruments to test air and document industry processes. As Frank Finan, a member of Breathe Easy, noted on the reasons for monitoring air, "We decided on air. It affects everybody."

Although residents and community groups had undertaken water-quality monitoring to assess pollution from fracking, individual residents could have very different exposures to water pollution. Some private wells could be contaminated while others were not, and some residents might obtain their water from municipal supplies. In contrast, air pollution was a more pervasive problem, yet it too was unevenly distributed in the community. Those who lived downwind of compressor stations might suffer much worse air quality than those who have not yet had infrastructure encroach on their home or work environments. Nevertheless,

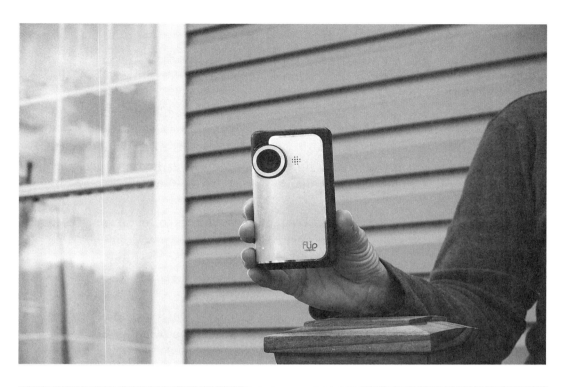

Figure 2.10. Participant showing use of a Flip video camera for documenting increase in traffic due to the fracking industry and transport of equipment. Photographs by Citizen Sense.

air provided a "focus," since as Frank explained, the group sought "to be focused, to forget about every other aspect of our lives that got screwed." For Frank, this focus on air involved buying and using monitoring toolkits. He bought a toolkit to test water and air as well as radon. He purchased a "four-gas sniffer," and many other instruments that he noted required considerable labor to learn about and use. Because of the effort in using these different instruments, he decided to focus on the "gas finder camera" (the FLIR introduced earlier in this chapter), along with photography, to document the effects of the fracking industry and poor air quality.

Some of these monitoring practices required residents to collect samples for lab analysis. For this reason, citizen-sensing practices that produced "real-time" data on air-pollution levels generated considerable interest as a way to expand ongoing environmental monitoring practices. By comparison, while buckets and similar techniques for monitoring air pollution have been used for fence-line monitoring at refinery sites and as part of environmental justice campaigns, buckets do not generate real-time data.[44] Sensors could provide a more immediate picture of environmental conditions. However, they do not lead to a direct trajectory from data collection to environmental action as change. As it turns out, many complications arise when citizens collect data about air quality.

The gathering of "evidence," which monitoring technologies initially seem to enable, raises more questions about how monitoring is undertaken, how data are collected and managed, how to translate the data into policy and action, and how practices for sensing pollution could expand potential infrastructures of care. Citizen-sensing practices collect data about particular pollutants. At the same time, they can attend to parallel "qualitative" data such as noxious smells, noise, and health effects, which in turn can shift the categories and procedures for how evidence forms. Yet these diverse data types can be challenging to mobilize for political change. They might not easily align with or circulate within regulatory frameworks.

As Michelle Murphy has suggested in her comparison of toxicology tests to citizen-led monitoring practices of indoor air pollution, these diverging evidentiary practices can make present or "perceptible" different aspects of chemical exposure.[45] Toxicology tests focus on how individual chemical concentrations create distinct bodily effects. Yet these practices might not register the diffuse and multiple modes of exposure that are difficult to describe within singular and causal dynamics. By comparison, citizen-led health studies could present a more situated and lived experience of chemical exposure. By registering lived experiences of chemical exposure, such citizen practices could "instigate" other forms of political action, even if they do not align with regulatory frameworks.[46] The

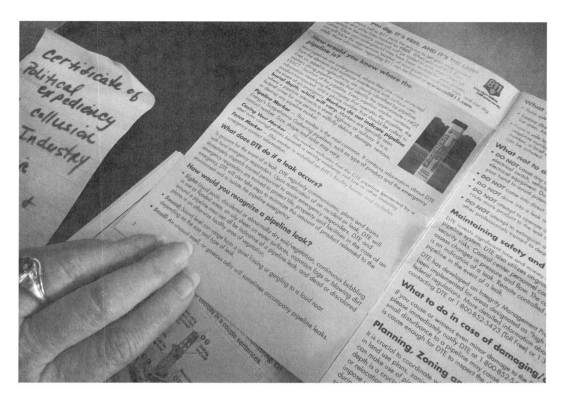

Figure 2.11. Participant showing different brochures and guidebooks for detecting and reporting a pipeline leak and for monitoring radon. Photographs by Citizen Sense.

Figure 2.12. Participant showing decibel meter for documenting noise from fracking industry and showing photo album documenting the changing landscape. Photographs by Citizen Sense.

process of making pollution present and sensible could be differently approached as concentrations, experiences, and lived encounters that anticipate harm to environments, bodies, and politics. At the same time, future effects could evade present perceptibility.[47] Speculative citizenships materialize through this anticipation of indeterminate future effects that mobilize current practices to document, analyze, and struggle toward more breathable worlds.

COLLECTIVE ENVIRONMENTAL INQUIRY

While reviewing monitoring practices already in use within this particular community affected by the fracking industry, the Citizen Sense research group further engaged in participatory and practice-based research to build, install, and test sensor technologies. These mostly digital devices could generate real-time data. They offered different ways of documenting harm and, potentially, of acting on polluting conditions. Through this approach, we then worked with communities to understand how monitoring practices emerge and change as they attempt to account for lived experiences of energy extraction.

When undertaking this research, the Citizen Sense project worked with residents to develop a monitoring toolkit that could monitor air pollution in everyday settings. As part of the collaborative aspect of the research, we established a dialogue with residents of northeastern Pennsylvania about which pollutants and environmental disturbances they were already in the process of monitoring. We studied how and why they undertook environmental monitoring practices, what wider networks were important for communicating their findings, and how it might be possible to work together to develop a citizen-sensing toolkit that would be useful for monitoring air pollution from the fracking industry.

Through a back-and-forth exchange that included several in situ meetings and remote teleconferences with residents, we developed a "Logbook of Monitoring Practices." As detailed in the previous chapter, this was a preliminary toolkit for participants to document their existing monitoring practices, note their particular observations and concerns about how fracking was changing landscapes, and indicate who should be monitoring and what should be monitored. We collected nearly thirty of these completed logbooks. Based on the logbook entries, along with images and video submitted by residents documenting their environments, we identified several possible monitoring technologies and practices that we began to assemble into a Citizen Sense Toolkit for use and testing.

We then developed a Citizen Sense Toolkit for monitoring fracking-related air pollution over several months, spanning from autumn 2013 to the summer of 2014. We designed the kit through ongoing discussions with residents and participants about the primary pollutants of concern, from nitrogen oxides to particulate matter and noise. Environmental science research that university researchers had conducted in the area also informed people's interest in monitoring particular pollutants. Nitrogen oxides from compressor stations were of concern due to continuous emissions and blowdowns. Nitrogen oxides are also criteria pollutants[48] and can indicate ozone formation. Methane was a pollutant of interest because it could offer a way to detect leaking gas across multiple sites within the fracking infrastructure. While not an air pollutant per se, noise was a topic of considerable discussion, since many people experienced disturbed sleep from the noise and vibration of infrastructure. Particulate matter, which was emitted from diesel trucks and generators and multiple other sources, was of concern as a pollutant particularly hazardous to human health. And VOCs from BTEX to glutaraldehyde were discussed as pollutants specific to petroleum and fracking production that could be monitored to indicate emissions from these industries.

Through research into which sensor technologies might be most adaptable, affordable, and accessible over a longer period of use, we then developed and assembled a Citizen Sense Toolkit of multiple components that were off-the-shelf or developed specifically for the monitoring situation. The toolkit included a Speck $PM_{2.5}$ digital monitor, which sensed, displayed, and recorded particulate matter levels in real time; industrial analog badges, which passively sampled air and monitored personal exposure to BTEX compounds; and several custom-made Frackboxes developed by Citizen Sense, which were placed next to compressor stations and monitored nitrogen oxide, nitrogen dioxide, ozone, and VOCs, as well as temperature, humidity, and wind speed.[49]

The Citizen Sense Toolkit was an assemblage of newly developed instruments as well as existing sensors. We borrowed the Speck monitor from the Create Lab at Carnegie Mellon University, which was making its device widely available for environmental and health groups to use throughout the state. Create Lab was distributing Speck sensors at public libraries in Pennsylvania, loaning devices to environmental health groups, and donating monitors to communities to use in their local areas.[50]

Along with the Speck monitor, we tested and developed our own Frackbox air-quality monitors that housed sensors and a weather station in jumbo black-steel post-mounted US mailboxes. Designed to blend into the rural landscape, the Frackboxes included sensors for monitoring ozone, nitrogen oxides, and VOCs, along with temperature, humidity, and wind direction. A prototype technology, the Frackbox used newly emerging sensors from Alphasense, a research group and factory in Essex, UK, which was developing low-cost air-quality sensors. In parallel to these digital devices, we also included analog BTEX badges from health and

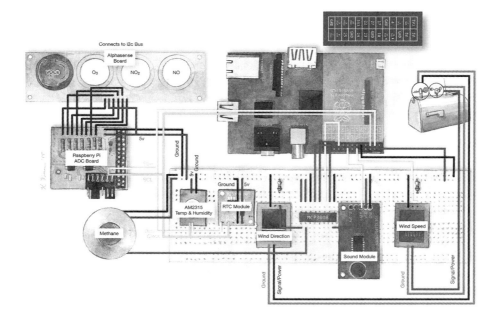

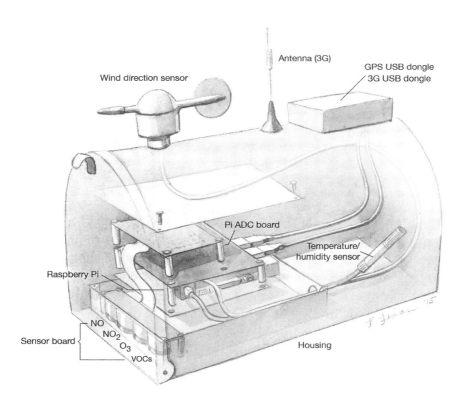

Figure 2.13. Diagrams of Frackbox wiring and Frackbox components. Illustrations by Kelly Finan; courtesy of Citizen Sense.

safety suppliers to test and compare monitoring techniques that local residents had used. These multiple sensors formed the kit of parts that became the Citizen Sense Toolkit and were differently bundled into networks of technology, research, health, communities, and infrastructure in ways that informed the practices and circulation of these devices and their data.

In addition to these different air-quality sensors, the toolkit included a custom online platform. The platform located, logged, and displayed environmental sensor data so that participants could access data and see how the community-monitoring network was forming in relation to sites of concern. Because people felt the monitoring activities could uncover sensitive or controversial findings, we set up the platform as a private site only accessible to participants during the monitoring period to ensure that monitoring locations were not disclosed. Together with the platform and sensors, the Citizen Sense Toolkit included a second logbook, which consisted of a how-to guide and instructions for using the various components of the toolkit. The logbook and online platform provided space for recording observations related to health effects, changes in the environment, and industry activity underway that could be used to explain patterns in citizen data sets.

The Citizen Sense Toolkit was assembled through a process of collectively asking how-to: how to monitor pollutants, how to develop or source sensors, how to site monitoring equipment, and how to record data. Each of these points of collective inquiry opened into discussions of previous research on pollutants and contributors' experience with monitoring, as well as knowledge of instruments and ways of collecting and presenting data as evidence. The how-to aspects of developing an environmental sensing toolkit formed through practices of making and testing while also drawing on earlier monitoring studies and experiences.

The tension between following set protocols for monitoring—which in many cases were not fully established for citizen monitoring—and working in a more experimental register became a dynamic that was collectively yet differently negotiated to make room for the more prospective aspects of these open-air toolkits. The point was not to replicate findings from research that scientists might have carried out previously in the area, but rather to study and document the lived experiences of residents who brought multiple insights to bear on the problem of air pollution and fracking. At the same time, participants were keen to generate "hard data" that would be taken seriously by regulators and ensure that their experiences of harm would be taken into account and addressed. Here, data collection becomes continuous with anticipation, if not speculation, of how to organize to address pollution. As part of this how-to mode of inquiry, not only is the separation between observing and acting undone,[51] but also modes of action

Figure 2.14. Specks set up for a workshop and loaning to participants. Photograph by Citizen Sense.

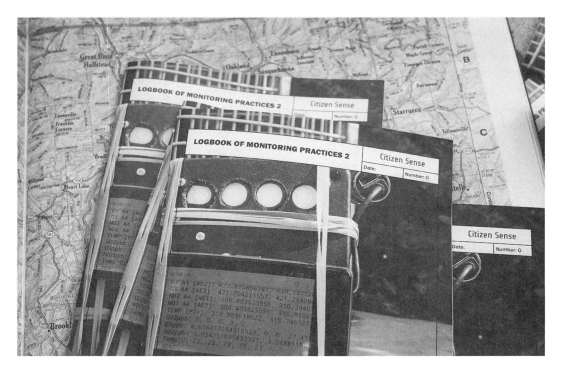

Figure 2.15. Second logbook developed by Citizen Sense, providing instructions for participants for air-quality sensors in the Citizen Sense Toolkit. Photograph by Citizen Sense.

influence the relations that empirical research is meant to have—or might have, in a more speculative register. Here is a method of radical empiricism aligned with open-air instrumentalism, where citizen-sensing practices generate and propose distinct relations, practices, and engagements that work toward more breathable worlds.

How to Activate Open-Air Inquiry

As part of the process for distributing toolkits, in October 2014 the Citizen Sense research group hosted a series of events in northeastern Pennsylvania. The events included a community workshop where the Citizen Sense Toolkit was introduced, a walk along fracking infrastructure on which various monitoring equipment was tested, and a roundtable to discuss broader issues related to fracking and community organizing.[52] We invited community members along with speakers who had experience with environmental monitoring, public health, and fracking to contribute. Participants included residents, technologists, environmental-health practitioners, local ecologists, and community organizers.

During the workshop, we distributed the Citizen Sense Toolkit to residents to test, take home, and install. But even more than learning about the monitoring toolkit, the workshop created a forum for the how-to. This was a space for working through and asking how to monitor air pollution from fracking, what to monitor and where, how to analyze and communicate data, and how to connect findings to other experiences such as health effects. While the workshop sought the input and experience of all involved who might help to identify problems and discuss ways of monitoring, the gathering was one where many people in the room were not necessarily on speaking terms, since the pressures and strains that fracking had placed on the community had led to lasting rifts between people. Monitoring and data collection were perceived to offer a possible neutral zone. Infrastructure could be studied to establish whether it was contributing to elevated pollution levels, and actions could be taken based on evidence. Neutrality was a movable condition that relied partly on our role as "third-party" researchers external to the community and partly on the role of sensors and data as the evidentiary techniques that could be used to hold industry to account.

In this sense, the workshop was also a chance to air concerns and develop a more collective approach to inquiry. As part of this process of inquiry, our research group communicated what our project motivations were for studying pollution in the area, since this information helped residents to understand what our commitments were and decide whether they felt the research was organized to facilitate their own questions.[53] People wanted to know whether we were funded by industry and our views on oil and gas extraction. We explained how the project was studying the rise of citizen-sensing technologies and practices, and how or whether these were contributing to new or more effective approaches to documenting environmental concerns. We also noted we were interested to understand what practices and questions emerge when people undertake monitoring and use data to address environmental pollution, which we also hoped to address.

This more speculative aspect of the research relied on a collaborative approach to working with the sensors to see how they could contribute to different forms of environmental engagement. We noted that monitoring data could be used in many different ways to document environmental concerns. At the same time, monitoring data alone might not be sufficient to address problems and regulators or industry might not accept citizen data. Documentations of experience alongside sensor data could, in this sense, begin to develop a more compelling narrative about air pollution in the gas fields. We also emphasized that the research process involved open development, where neither the technology nor the research design was entirely "finished" and that the project would continue to take shape as the sensors were installed and used. As a process of open-air instrumentalism, this research was at once speculative and collaborative. It attempted to undertake collective inquiry to respond to changeable conditions while suspending the rush toward specific outcomes. Such an approach could allow other kinds of inquiry—or science—to emerge.[54] This speculative process of open-air instrumentalisms sought to understand how sociotechnical engagements unfold in worlds, as they also work to make breathable worlds.

As part of the citizen-monitoring project launch events, we then took monitoring toolkits out on a drive and walk to infrastructure sites to test monitoring practices and technologies. The drive–walk allowed us to discuss issues related to fracking as well as how best to monitor in particular settings. The walk became an extension of the workshop as inquiry, and yet here we literally moved into the open air. We tested an array of monitoring equipment while also experiencing numerous infrastructural sites and the distinct patterns of pollution that they created.[55]

We began our outing as a group by first driving to the Tennessee Transmission Pipeline. We undertook a walk to see the patterns of forest clearing, land grading, and pipeline installation that characterize this part of the natural-gas infrastructure. A local ecologist, Nancy Wottrich, explained the effects pipelines had in carving up larger ecologies into more fragmented spaces; she noted that this often led to reductions in biodiversity, since many organisms required larger intact ecosystems to survive. Next, we walked from the pipeline to a nearby compressor station, where we had gained permission from the landowner to install

Figure 2.16.
Gas pipeline
infrastructure
encountered during
walk with
participants.
Photograph by
Citizen Sense.

a Frackbox to monitor emissions. As we walked to the compressor station with
handheld methane monitors, ultrafine particle detectors, badges, and several other
monitoring devices, we detected a palpable and acrid wave of air. The churning
of the compressor station made it difficult to talk above the noise, and Nancy
mentioned that noise was also a pollutant that could damage local ecologies and
organisms.

As a mode of open-air inquiry, the walk moved the experience of fracking
infrastructure to a more central if even debilitating aspect. We decamped to the
road to discuss further how best to monitor and record emissions from these
sites. We then drove to our next stop, a gathering line and well pad where active

construction was underway. We first looped around the well pad by climbing an adjacent hill and overlooking the construction site. Diggers and trucks mechanically scraped, leveled, and hauled dirt to create an expanded well pad location, where additional wellheads were being added to the site. Erosion fences were in place next to an extensive area of land that had been seeded after clearing. However, the seed mixes mostly consisted of fescue and clover, a greatly reduced palette that Nancy reminded us bore little resemblance to landscapes prior to disturbance from fracking. Here, new fracking ecologies were being shaped, affecting air, water, soil, plants, animals, and people.

We turned back down to a dirt road near an adjacent gathering line. At this stop, we heard from Laurie Barr, a resident of Pennsylvania who had started a project for monitoring lost, abandoned, and orphaned wells. Because Pennsylvania has been the site of extensive activity from the extractive industries, there are also numerous leftover pipelines, abandoned wellheads, and leaking infrastructural components that continue to affect environments. With her citizen-led project, Laurie had begun an initiative to document, map, and monitor these lost, abandoned, and orphaned sites. She passed around maps of wellheads and showed night-vision photographs of deer drinking briny water at leaking wellhead sites. She also showed her "gas finder" monitor for detecting methane, which she used to assess whether gas leaks might occur at these sites.

As a how-to mode of inquiry, the walk visited infrastructure to observe and document industry operations underway. We observed a Frackbox installation next to a compressor station, investigated a new well site under construction, and listened to community organizers describe their own practices of undertaking environmental monitoring and gathering data. We spoke to residents living near infrastructure to learn more about their day-to-day experiences of industry operations. We also heard about attempts to work with state and federal regulators to understand existing monitoring infrastructures and environmental data, and how these could address fracking-related pollution.

The walk offered a chance to investigate ecological disturbances and learn how new gas-field ecologies form through linear excavations carved into soil and forests, along with grading and clearing that reshaped environments for wellheads and compressor stations. In this way, the walk formed a collective experience of studying these infrastructural ecologies. It also made palpable the unequal experiences of harm from fracking. Some people lived surrounded by compressor stations, and others had second homes in the area. As researchers, we were primarily located in London, far from fracking but not removed from the problem of air pollution. We added to the inventory of harms experienced through the walk and conversations. And we considered how to document pollution with monitors, diaries, data, platforms, social media, and community organizing as part of the more extensive proposal for how to work toward more breathable worlds.

The walk, workshop, and roundtable served as collective modes of inquiry as well as forums for discussing the problems of air pollution. The concerns and proposals that materialized informed the next steps of the monitoring process, which involved setting up components of the Citizen Sense Toolkit in monitoring locations. How and where to set up sensors, how to monitor and for how long, how to ensure the data would be useful and be listened to: these were all recurring topics in our multi-sited and multi-day conversations as we installed sensors. We visited residents, often at their homes, to help set up devices, ensure that connections were made to the data platform, and discuss issues related to monitoring. Participants were interested in monitoring at several sites, comparing infrastructural locations, and even surrounding infrastructure with sensors. While we had a limited number of monitors, we began the process of visiting locations of concern, working with participants to set up sensors, and establishing a connection between sensors and our platform so that participants could view data in real time and over time, as well as compare their data to other monitoring locations in the network.

Numerous questions arose in the process of setting up Specks, which was often far from straightforward. Sheltered outdoor locations needed to be identified, power cables needed to be sourced, duct tape had to be procured, and Wi-Fi had to be connected. The minor digital infrastructures that enabled monitoring became sites to identify, adapt, and stabilize to undertake monitoring. The online platform similarly required tussling with home PCs and ancient operating systems, Internet Explorer browsers, and multiple components that were not part of our original testing of the website. Eventually, multiple monitoring locations came online and a community-monitoring network began to take shape and grow, surrounding infrastructure, roadways, and homes in this three-county area of Pennsylvania.

Along with the Speck monitor setup, we placed Frackboxes at three strategic locations next to compressor stations to monitor this industry infrastructure that was of particular concern. Setting up the Frackboxes required installing mailbox posts (in some cases) or sourcing stands, along with power for the Frackbox to function. One Frackbox was powered by solar energy, and this required setting up the PVC panel and battery in a plastic tub in the woods. Data were piped over a 3G dongle to the internet, forming an at times precarious connection in this remote location. While they were a provisional and test device, the Frackboxes and their data were of considerable interest in the community. In the early stages of setup, Fox News learned of the devices through a community member and

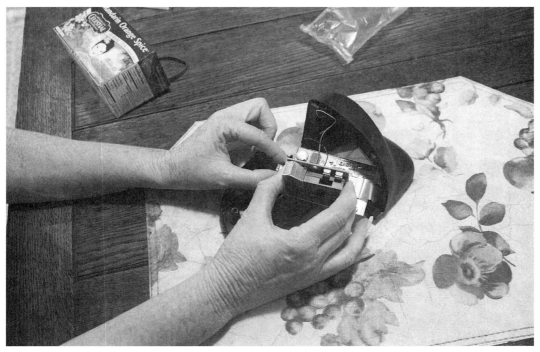

Figure 2.17. Setting up Specks for monitoring particulate matter at participants' monitoring locations. Photographs by Citizen Sense.

contacted us about the data. We told the reporter that the locations were confidential and that the data had yet to be analyzed or verified. Nevertheless, a cat-and-mouse game ensued, with one Citizen Sense researcher having to duck out of a hotel early in the morning when learning of the reporter's plan to intercept the researcher in the hotel parking lot. Working with technologies in the open air, we learned, was also a process that could lead to overlapping and controversial forms and worlds of inquiry.

How to Transform Data into Evidence

In all, we distributed nearly thirty Citizen Sense Toolkits to participants, which they used over a period of seven months. We also installed three Frackboxes that ran for the duration of the study, even in the depths of winter during blizzard conditions. We had initially planned to monitor for three months, but new participants continued to join. A broad range of monitoring practices materialized across multiple locations. Some participants produced long-term, continuous data sets while recording their experience of pollution. Others contributed data for a few months or weeks until devices had to be moved or unplugged. Participants logged their approximate monitoring location on the Citizen Sense Toolkit platform, primarily through the use of the Speck $PM_{2.5}$ monitor, but also by logging their observations of industry and other activity that might generate high $PM_{2.5}$ levels. Data were available to view both in real time on actual Speck devices and on the platform once uploaded. Observations and readings could be compared across distinct monitoring locations, and in some cases discussion arose about the multiple readings, techniques, and events that might cause elevated pollution levels. Participants considered whether nearby gas infrastructure caused elevated readings, or if high pollen counts or other industries nearby were causing spikes in their data.

This collaborative process unfolded an ongoing set of questions about how to monitor, what to do with their data once it was collected, and how to ensure regulators take their data seriously. At various points, participants suggested that their data were not indicating anything of significance, which often meant that their lived experiences of odor or nuisance or perceived emissions did not match up with real-time displays on the Speck or in the data collected and available on the platform. There was a sense that an immediate register of harm should be evident, or else the device was failing to perform. Here, the device did not make pollution evident in the way it was expected to. The sense that pollution must be present created dissonance across machinic and bodily experiences of environmental events.

Figure 2.18. Setting up Specks for monitoring particulate matter at participants' monitoring locations, and analyzing citizen data using the Airsift 1.0 platform. Photographs by Citizen Sense.

On other occasions, Speck monitors would provide high readings, or spikes in $PM_{2.5}$ levels. There was often a process of troubleshooting to understand what could be causing the high readings: Was it a device malfunction, or was a pollution episode or some other atmospheric event underway? On one occasion, two participants and members of Breathe Easy, Chuck and Janis Winschuh, called the Pennsylvania DEP in order to lodge a complaint in relation to a high $PM_{2.5}$ reading on their Speck. Chuck and Janis found that when they were visited by the DEP, industry representatives also came to their home to find out what monitoring equipment was in use and how the study was organized. The participants' concern was that neither the DEP nor industry actually attended to the high readings they recorded, which occurred over several hours and were not due to a faulty device.

Chuck and Janis subsequently took their story to the media, which documented how regulators and industry responded to their citizen-sensing activities and concerns about pollution.[56] In the process of attempting to evidence harm, Chuck and Janis found that their data were of less interest than the act of citizen monitoring. Here, the "evidence" of harm materialized not through data sets but through participating in a citizen-sensing study and having that participation queried. Such an arrangement resonates with Murphy's suggestion that monitoring can be as instigatory as it is evidentiary, since the DEP and industry were apparently more attentive to the "fact" of community organizing than to the data they were collecting.

Yet this is by no means to discount the importance of data collected by citizens, since it has also been used to document pollution levels of concern. Other participants, Meryl Solar and Rebecca Roter, found that the intense and rich data sets collected, which numbered over five million data points by the end of the monitoring period, could be mined for patterns using Microsoft Excel spreadsheets to indicate that harmful levels of $PM_{2.5}$ were occurring at several sites across the community-monitoring network. Meryl and Rebecca used these data to arrange a teleconference with the Pennsylvania DEP, the Pennsylvania Department of Health, the federal Agency for Toxic Substances and Disease Registry (ATSDR), and the Centers for Disease Control, along with Citizen Sense, to discuss their findings. Agencies and regulators were skeptical about the citizen-sensing devices and raised queries about their calibration and use as well as the validity of the data. Yet Meryl and Rebecca were able to use a combination of data and experience of lived exposure to make a case for regulators to undertake follow-up monitoring at one of their homes.

While we were analyzing and communicating preliminary findings from the data, it became clear that the graphs available on our provisional platform could not fully convey the different patterns of pollution experienced. How to analyze data then became part of the long list of how-to items that unfolded through this collaborative research. While working with a King's College London atmospheric scientist, Benjamin Barratt, we learned of an open-source software, openair,[57] which we could use and adapt to analyze citizen data. With the objective of making a DIY data-analysis platform, we then made a citizen data toolkit, Airsift, which would allow participants to analyze their data more fully according to the time of pollution events, the direction of pollution, and the likely source of pollution based on wind speed, humidity, and a host of other variables. We were assembling an expanding community-monitoring infrastructure that adapted to conditions taking place quite literally in the "open air," in response to political engagements and pollution events. These open-air instrumentalisms did not arrive at a finished condition as such but rather worked through ongoing processes of inquiry to generate practices for evidencing harm. They were, in this sense, forming speculative citizenships that struggled to create tactics and forums for reducing and mitigating environmentally destructive practices.

With the plots and graphs we were able to generate from the Airsift toolkit, we worked with participants to collect on-the-ground observations and experiences of pollution that together formed narratives for five key locations in the network. We developed these narratives into data stories according to the township where the monitoring had taken place.[58] The data stories documented pollution levels in Bridgewater, Brooklyn, Dimock, Mehoopany, and Liberty Townships. The data stories analyzed the citizen data and observations of pollution events, while providing indications for how to mitigate or reduce pollution based on findings. Citizen Sense launched the Pennsylvania Data Stories in April 2016 and, along with participants, shared the findings with state and federal regulators.[59]

Soon after we launched the data stories, the ATSDR released a report that documented the results from their parallel follow-up monitoring.[60] Their report documents how they found elevated $PM_{2.5}$ levels at the test monitoring location, and higher pollution levels were likely attributable to nearby infrastructure. The report corroborated the citizen data findings at this Brooklyn Township location. These findings also led the ATSDR to recommend that the Pennsylvania DEP develop more robust practices for monitoring and mitigating emissions, particularly from industry.

Just after the ATSDR made its report public, the Pennsylvania DEP announced that it was undertaking an "unprecedented expansion" of its $PM_{2.5}$ monitoring network.[61] In turn, the fracking operator whose particular infrastructure was near the ATSDR monitoring location responded that it was disappointed by the DEP's decision to undertake additional air-quality monitoring and that it found the ATSDR's report to be based on "speculative" data.[62] Commenting on this news, Rebecca indicated that a speculative approach was in fact not a bad thing, since waiting for harm to be done and then conducting "retrospective public health studies" was less advisable than taking action before harm was done.

The process of evidencing harm drew on multiple forms of data and evidence, some of which could be considered "speculative" exactly because they were generated through provisional practices. Yet these practices enabled residents and agencies to make a case for greater levels of care in the form of monitoring and attending to exposure from fracking to avoid harm before it occurs. Speculative monitoring practices became a way to evidence and work toward reducing harm. Rather than wait for harm to be done, the citizen-sensing practices demonstrated that pollution was occurring at elevated levels and that action should be taken to prevent future harm. Speculative citizenships materialized here as a practice for proposing political subjects and engagements that could work and struggle toward more breathable worlds. Rather than simply document pollution once it has affected environments and health, here speculative citizenships sought to prevent pollution before it occurs.

In 2018 the Pennsylvania DEP installed a regulatory air-quality monitor for $PM_{2.5}$, carbonyls, and VOCS in an area where citizen sensing had taken place.[63] Situated in the township of New Milford, the monitor signals to the community that the DEP has begun to take their concerns about air quality more seriously.[64] However, the expansion of this network also raises questions about the extent to which monitoring can become a practice of preventing harm, or whether it can become a mechanism for allowing an "acceptable" level of pollution to occur. Monitoring practices could create an uneven relationship to improving or detoxifying environmental conditions, since polluting industries could potentially expand if an official monitoring location can demonstrate that pollution levels do not exceed regulatory guidelines. Moreover, under a new administration, the federal EPA dialed back regulations for oil and gas, in a move that suggests evidence and harm can be the least of concerns where extractive industries are concerned.[65]

The overall inquiry of "how to evidence harm" that guides this collaborative research is how or whether evidence could be sufficient to address or forestall harm. Here, somewhat remarkably, citizen data contributed to evidence-based policy that led to the expansion of a state air-quality monitoring network. And yet, the influence that citizen data had in this process and decision was not always publicly acknowledged.[66] Evidence and recognition of harm were not necessarily communicated through the more accountable and transparent channels and forums of governance and industry engagement that communities sought.

Moreover, not all residents who undertook monitoring would necessarily have equal access to creating and communicating evidence. For instance, property ownership might seem to confer more rights on people who monitor air quality than on those who do not own property. These forms of speculative citizenship might in turn reinforce, rather than transform, unequal forms of property-based citizenship formed through economic inequality and settler colonialism. At the same time, it is a curious contradiction that property ownership does not confer as many rights as might be expected, where eminent domain is frequently exercised to install pipelines and infrastructure across the gas fields. In a related way, homeowners in Flint, Michigan, did not seem to have greater sway over the problem of water pollution.

While certain evidence can count in distinct situations when communicated by well-connected actors, it might also be overlooked or ignored, depending upon prevailing political interests and socioeconomic power relations. In other words, the drive toward creative democracy and inquiry that would work through a Deweyan critical and engaged intelligence is thwarted by power struggles and inequalities. As Cornel West writes in his study of pragmatism, such striving toward radical democracy can present severe limits for "the wretched of the earth, namely, the majority of humanity who own no property or wealth, participate in no democratic arrangements, and whose individualities are crushed by hard labor and harsh living conditions."[67] Rather than unfolding as participatory democratic exchanges, practices of gathering and presenting evidence could reinforce and exacerbate inequalities. Indeed, regulators and industry could issue a repeated demand for evidence—to prove that pollution is occurring—but then disregard that evidence.[68] The gathering of evidence could then lead to exhaustion and injustice, a topic that will be taken up and further expanded upon in the next chapter. Yet here it is important to note the variable and uneven ways speculative citizenships can materialize when attempting to evidence harm and generate responsive practices of care.

How harm registers, the forms of evidence that are admissible, the subjects that can convey and act on evidence, and the worlds that are varyingly configured or denied—all of these are expressions of power. How to evidence harm asks how it could be possible to account for more and other experiences in struggles for more livable and just environments. Speculative methods, as Ruha Benjamin notes, can be a way to extend social practices toward possibilities for greater equity in the facts that are taken into account and how they are acted upon. Working in a register of speculative fiction, Benjamin suggests that these modes of inquiry can offer a way to "experiment with different scenarios, trajectories, and reversals, elaborating new values and testing different possibilities for creating more

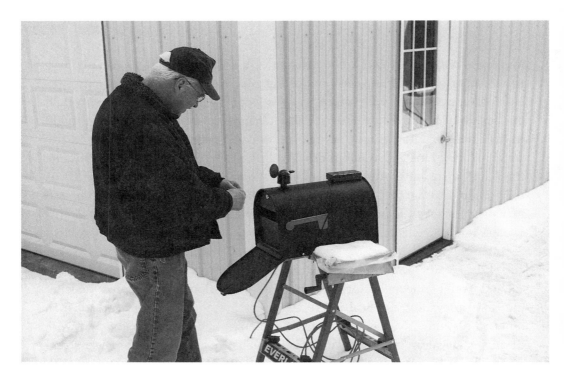

Figure 2.19. Frackbox installations at two different locations near compressor stations and other fracking infrastructure, including a toolkit hosted by Paul Karpich in Dimock. Photographs by Citizen Sense.

just and equitable societies."[69] As processes of open-air instrumentalism and collective inquiry, citizen-sensing practices demonstrate that evidencing harm is a complex process that organizes sensing, facts, relations, environments, and speculative citizens toward different possible environmental inhabitations. How-to as a process of evidencing harm can designate a speculative mode of inquiry. How-to can also demonstrate the politics whereby decisions are made, as well as propose how to struggle toward and realize more breathable worlds.

HOW TO EVIDENCE HARM

In this account of working with residents in the gas lands of northeastern Pennsylvania to monitor air pollution, it becomes clear that citizen-sensing toolkits materialize as much more than digital gadgets or makerly gear. By tackling concrete problems, citizen-sensing practices and technologies quickly become bound up with environments, communities, institutions, and wider politics. The accuracy of monitoring devices, the monitoring protocols used, the legitimacy of the data, and the agendas of communities all influence citizen-sensing practices and citizen data. Here, participation involves much more than merely using a sensor to collect data about a particular pollutant. Instead, it sprawls into struggles for how to be and become citizens, and how to make more breathable worlds.

This discussion of collaborative air-pollution research examines how citizen-sensing practices generate alternative forms of evidence to document harm that is often overlooked or neglected. Citizen sensing here moves beyond the narrow outline of "data to action" to open into distinct worlds of inquiry and political struggle. Different relations and communities might be activated through monitoring practices, or existing communities might reencounter persistent problems while finding ways to hold environmental regulators and industry to account. Moreover, the citizen-sensing data do not always readily circulate to relevant agencies. Instead, data are potentially generated in excess, difficult to collate and present, and subject to disputes about their legitimacy.

Attempts to "care about air" generate speculative citizens and worlds in the making. Multiple practices and infrastructures of care could materialize as speculative approaches for evidencing harm. These practices even become necessary to express how harm materializes outside of or in the absence of protocols and practices recognized by environmental regulation and policy. Speculative forms of citizen-led environmental sensing could facilitate the process of generating new approaches to what counts as evidence to include registers of experience that might ordinarily be dismissed or overlooked. By opening up air-pollution monitoring to these expanded approaches to data and evidence, it could then be possible to incorporate speculative engagements not as the opposite of evidence and "proof" but rather as an indication of how citizens are demonstrating what matters to them in their lived environments and how they are attempting to bring their experiences into spaces of recognition and relevance.

Sensor technologies often promise an ease of participation and contribution to environmental problems. Such promises could be tested and even critiqued. Yet sensing technologies can also give rise to struggles to democratize environmental monitoring and evidence. New forms of environmental politics and expanded approaches to capacity building could be generated through these efforts to address environmental problems. At the same time, a scientific approach to encountering environmental issues—establishing a hypothesis, evidencing this with data, and bringing forward findings—does not necessarily fit so neatly with a potentially more distributed, community-driven, qualitative as well as data-based set of concerns about the environmental effects of fracking. If emphasis is primarily placed on gathering data to evidence claims, then other modes of organizing might be less foregrounded, even though they are crucial to developing collective approaches to environmental problems.

The Citizen Sense project has been committed to investigating community-monitoring practices already underway as well as rethinking and reworking what monitoring practices might become through practice-based research. Research into environmental monitoring could, in this sense, attend to how diverse modes of evidencing harm are generative of collective practices of care. In relation to social research and practice, this could generate distinct approaches to engaging with environmental communities, speculative citizenship, and participation—as collective inquiry and struggle. These are undertakings that often proceed from more apparently epistemic and information-based starting points: how to gather evidence to demonstrate the facts of pollution. And yet, what might it mean to undertake an environmental monitoring project from the perspective of *experience* as a form of evidence in the making, and not just from information and awareness? Such a question, as I have discussed in relation to citizen sensing and citizen data, is concerned not simply with how facts take hold, but more centrally attends to how experience is a critical part of speculative propositions and their effects that might generate more breathable worlds.

In this sense, I understand speculative practices for evidencing harm to offer up as much an opportunity as a dilemma, a challenge as a creative opening, since these sensing practices might generate more accounting-based ways of understanding environmental problems by limiting speculation. In other words, they could document pollution without providing any clear indication of how to act. Alternatively, citizen-led monitoring could generate open-air instrumentalisms and speculative configurations for addressing situated environmental concerns.[70] Practices for evidencing harm could then anticipate and speculate toward ways of addressing harm through attending to lived experiences of environmental destruction.

As discussed in the previous chapter, this is a more propositional approach to evidence, or as Dewey has suggested, a focus on "consequent phenomena" rather than "precedents" that attends to the "possibilities of action" and the constructive functioning of thought.[71] West refers to this approach as a "prospective instrumentalist viewpoint."[72] Open-air instrumentalism outlines this prospective approach to evidence, where ways of observing and experiencing worlds also constitute potential courses of action. Open-air instrumentalism encompasses such collective and multi-agential modes of experience: these actions exceed a willful liberal subject in their reliance on pluralistic relations and worlds of influence.

This discussion proposes that citizen sensing, when undertaken in a speculative register and through speculative trajectories of citizenship, could draw out the potential and instigatory—rather than simply descriptive—registers of these practices. If speculation is a practice generative of possible futures, then a speculative approach to evidencing pollution and harm from fracking could rework the problem of how to register fracking's impacts as well as work toward practices for mitigating emissions and exposure. Such an approach to researching environmental monitoring practices seeks simultaneously to engage with the more speculative aspects of monitoring as they are undertaken and to rework the subjects and potentialities of monitoring by adopting a more deliberately propositional approach to pollution sensing and to evidencing harm.

Environmental sensing and monitoring are practices of inquiry that set in motion speculative subjects and worlds. Speculative citizens form here as distinct political subjects and collectives through attempts to evidence and to prevent harm. They manifest less as predefined entities than as subjects that form by working through perceptive and affective problems in milieus.[73] Speculative citizens are thus less figures of belonging to a predefined territory in the usual sense of citizenship, and are more expressive of operations that contribute to the formation of politically engaged subjects. This way of parsing subjects is also not de facto human-oriented, since the actual entities of citizenship might form as conjugations of experience across sensors, data, toolkits, collectives, environments, pollution, air, and organisms. In other words, speculative citizens are designations of political pluralities and collectives. Speculative citizens are citizens of worlds, where both citizens and worlds are in the making within open-air sensing practices.

Figure 2.20. Frackbox installations through winter 2014 and summer 2015 at compressor station. Photographs by Citizen Sense.

Extending monitoring into a speculative register makes it possible to develop an account of the entities that are drawn together within pollution sensing to speculate about environmental events, politics, and futures. A speculative
and collective approach to pollution sensing could help articulate environmental
politics—and citizenship—differently. In other words, a speculative approach to
environmental monitoring could recast or reformulate the problem that monitoring is meant to address. Monitoring, as a speculative proposition, could in this
way be approached as an adventure not just in making things possible but also
in making things (and worlds) matter in particular ways. Monitoring expresses
a way of being for distinct worlds; it presents a proposition and its effects that
allow worlds to take hold. It articulates a "feeling for the datum" that issues from
ways of "possessing" a world.[74] Citizens, in this way, materialize along with care
and concern for worlds. Because propositions are generative of effects, Stengers
reminds us to attend to the question of what is required for any particular foothold to persist. In other words: "From what wager does your success proceed?"[75]
Such a question points to how particular commitments form worlds in which
speculative citizens, sensing practices, and breathable worlds come to matter.[76]

From this discussion of air-pollution monitoring, I suggest that it could be
possible to rethink care not as a prescriptive or normative relation but rather
as a speculative mode of encounter. Care materializes here through monitoring
practices that work to evidence experiences of harm to environments, health,
and breathable worlds. Such an approach further points to the importance of
adopting a deliberately speculative engagement with citizen-based monitoring,
since many experiences could have been overlooked, exposures could be undocumented, and harm could be yet to be understood. In this way, it might be
possible to approach monitoring as an evidentiary practice and as distributed
formations of experience. The "taking into account"[77] that monitoring puts into
play is more than a practice of producing a set of data on pollutant concentrations. Instead, this practice involves attending to how the speculative effects of
fracking register, whether through data, bodies, sensors, environments, water,
air, health, or political struggles. From this perspective, practices and policies for
"caring about your air" could shift both to address overall emission levels of
criteria pollutants and to consider the multiple ways in which exposure occurs,
is experienced, and continues to be generative of new practices and entities—
and harmful effects. Practices for acting on air pollution could then become as
speculative and responsive as the conditions they would address.

DUSTBOX TOOLKIT

Dustbox particulate-matter sensor and monitoring kit developed by Citizen Sense for monitoring air quality in Southeast London. Illustration by Sarah Garcin; photograph by Citizen Sense; courtesy of Citizen Sense. This toolkit can be found in a more extensive form online at https://manifold.umn.edu/projects/citizens-of-worlds/resource-collection/citizens-of -worlds-toolkits/resource/dustbox-logbook.

Chapter 3

DATA CITIZENS

How to Reinvent Rights

Air pollution occurs not just from petrochemical industries in rural sacrifice zones, but also accumulates and intensifies in cities. Diesel vehicles, the burning of fossil fuels, construction dust, industry discharges, and drifting agricultural emissions generate particulate matter, nitrogen oxides, ozone, volatile organic compounds, and sulfur dioxide. These pollutants cause and exacerbate conditions ranging from asthma to heart disease and stroke.[1] While cities worldwide suffer from poor air quality, pollution levels greatly vary across disparate sites. Air-pollution levels in London often exceed both World Health Organization guidelines and EU Air Quality Objectives.[2] Still, these exceedances are typically less extreme than air pollution experienced in major cities in Asia and Africa.[3] During occasional air-pollution events in Delhi, for instance, instruments have topped out at "999" and could not register further increases in pollution levels.[4] The environmental crisis of air pollution overwhelmed the devices and data used to measure and govern it.

The numbers that record pollution levels and mortality rates provide one way of assessing the problem of air pollution. Yet within these numbers, many stories often go undocumented about how pollution circulates, sediments, and accumulates in bodies and environments. Such toxic exchanges tend to concentrate in communities where people of color and low-income residents live. Air pollution is unevenly distributed and experienced.[5] The regulatory instruments and data that monitor and mitigate pollution are also sparsely located along the fault lines of environmental injustice.[6]

Although the official infrastructures and techniques for monitoring air pollution are meant to assure urban dwellers that constant monitoring, control, and even care are given to the air they breathe, ruptures in the systems and technologies of governance regularly occur. The expert data, technologies, and practices that would indicate that urban air is breathable become a target for questioning and frustration. Urban inhabitants at times doubt the accuracy of the air-quality data made public, or they rail against the inertia within urban and national governments that they feel do little to improve air quality even when monitoring data indicate that air is polluted. The urban-environmental burdens of air pollution and the unequal distribution of toxic air lead to challenges and disruptions to expert data and infrastructures.

In response to governance regimes that might be at turns inept or rigid, people take up low-cost and DIY air-quality monitors and apps to measure air pollution. Citizen-sensing practices are proliferating worldwide as people document pollution, assess their exposure to air pollution, adjust everyday routines, tackle polluting activities, and transform urban environments. Whether checking apps that collect data from citizen networks of sensors such as PurpleAir or installing or wearing sensors to track air quality in their immediate environments, people are building and referring to expanded and parallel monitoring infrastructures to create alternative ways of sensing and acting on air pollution. Citizen monitoring of air quality, and the citizen data it generates, become a way to document and respond to harmful environmental conditions. These practices express a right to breathable worlds.

Responses to air pollution form through a complex mix of regulatory monitoring networks, air-quality indices, mortality and morbidity statistics, public-health guidelines, citizen sensing, political protest, air-quality campaigns, home filtration systems, breathing technologies, low-emission transport routes, and policy proposals, along with international and local dynamics in the movements of air and pollutants. Within these multiple approaches to air pollution, this chapter examines how citizen-sensing practices of monitoring urban air pollution activate citizens as data citizens. In collaboration with the Citizen Sense project, residents, workers, and volunteers in the Deptford and New Cross neighborhoods of South East London took up sensing technologies to monitor air quality. They located citizen-sensing technologies adjacent to traffic corridors and construction sites where rapid urban development was underway to document pollution. And they worked with the findings from their data to attempt to intervene in and reshape processes of urbanization that were contributing to polluting conditions.

By discussing specific citizen data practices tuned to urban environmental change, I investigate how data citizens form through the collection and operationalization of data as a potential medium for democratic engagement. I investigate how sensor-based data practices constitute and activate data citizens. More specifically, I consider in what ways environmental monitoring practices and infrastructures mobilize rights—to data, air, and breathable worlds. Based on this approach, this chapter then examines how pluralistic data practices could circumvent and reinvent rights by making more breathable worlds.

This chapter investigates how practices of using digital sensor technologies to monitor air quality in South East London generate pluralistic and uneven formations of citizen data and data citizens. While data are often viewed as something collected *about* citizens—typically by large technology companies in the form of surveillance and tracking—this study describes how there are now just as many instances of data generated *by* citizens to address environmental problems. Whether to sense air pollution, narrate lived experiences through online platforms, challenge governmental readings, or document conflict in areas of development, people are collecting, analyzing, and acting on data to support and remake urban environments.

I suggest that when they use sensor technologies to collect data about air pollution, urban dwellers express a *right* to a certain standard of air quality. These practices activate the right to data and the right to clean air. Indeed, multiple rights potentially form through citizens' monitoring of air pollution, including the right to breathe, the right to monitor, the right to environment, the right to the city, the right to health, the right to data, the right to participate, the right to research, the right to be political, and even the right to experience. Some of these rights are established in law yet are not readily enforceable. For instance, the "right to breathe clean air"[7] is variously observed within some urban environments through regulations that establish a legal right to a certain standard of air quality.[8] At the same time, when exceedances of official standards occur, the process whereby these rights might be enforced can seem to be opaque and ineffective, even when legal challenges are mounted to ensure that pollution limits are observed.[9]

Rights are often integral to expressions of citizenship. Yet it can be somewhat unclear whether and how rights factor into emerging citizenships. These might be "new rights" that are not settled into law.[10] Such rights are in the making. They are sociopolitical formations that materialize through data practices. The right to data is not simply the right to collect and communicate a bundle of evidence, however. It also comprises the right to mobilize findings to provide different observations, challenge expert findings, and work toward more just and livable environments. Even more than providing evidence of pollution levels, however, documenting pollution expresses a demand to change the environmental conditions that cause pollution, from excessive traffic to constant construction and fossil-fuel consumption.

To complement the previous chapter's investigation of practices that evidence harm, this chapter considers how data practices form rights to breathable worlds.

People struggle to be in exchange with their milieus in ways that sustain them as citizens of worlds. Such practices express a right to constitute and be constituted by worlds in the making. In this sense, citizens are activated through political relations that compose worlds. Rights, moreover, are distributed within environments and infrastructures that mobilize and support distinct modes of political engagement and inhabitation.

Citizen data do not guarantee a remedy to the problems documented. Instead, data become a medium through which to figure worlds by monitoring, documenting, narrating, and analyzing conditions of disenfranchisement and dispossession.[11] Citizen data practices are likely to lead to additional struggles to address environmental pollution. They are contingent and ongoing encounters with urban political life. This chapter discusses different formations of data citizens, and considers how environmental sensing practices give rise to citizens, rights, and worlds in the making. I then document how the Citizen Sense project worked with community groups and residents in South East London to generate and collect data about urban air pollution. I explore how the expression of the *right-to* gives way to a multitude of *how-to* practices, including how to mobilize citizen-sensing infrastructures, how to figure citizen data, how to pluralize data practices, how to make urban worlds with citizen data, and how to reinvent rights. The citizen data generated through these air-quality monitoring practices express rights to monitor, inhabit, and cultivate less polluted environments, even when such rights are unevenly realized. As potential practices of combat breathing, they become a way to reinvent rights by working through concrete struggles not just to evidence harm, as discussed in the previous chapter, but also to build more breathable worlds that push against the constrictions of lived environments.

CONSTITUTING DATA CITIZENS

"Data citizen" is a term that is in widespread use across research, activism, and industry to describe how technopolitical actors are constituted through data practices. As discussed in the Introduction to this study, "citizen" is often applied as a democratic veneer to digital technologies. Within the tech industry, "data citizen" circulates as a term to suggest the relative accessibility of data technology and data-analysis techniques to nonexpert users. Rather than a political subject, here a citizen is rendered as an amateur who should have easier access to data and its devices.[12] While "data citizen" is variously deployed to refer to the intersection of data and subjects, it can be somewhat unclear how data contribute to the formation of citizens as political subjects. The processes whereby data constitute citizenship or enable political participation remain rather vague in this formulation of the data citizen. Citizens could materialize through practices of data collection and acting on political problems. They could also form through the necessarily political infrastructures of data collection and mobilization. In other words, citizens—and their possible collectivities—are programmed and formatted through distinct data practices.[13]

In one characterization, science and technology researchers Judith Gregory and Geoffrey Bowker suggest that data citizens are assembled through particular quantitative techniques such as wearable technologies. In their estimation, data citizens are constituted as distinct technological subjects with and through "an ecology of microdata."[14] Here, subjects with wearable technologies are not necessarily undertaking a deliberate plan of participation; instead, they form as data citizens through the ecologies they plug into. This is a very different characterization of citizenship, which forms through the conditions and relations of technological infrastructures.

Indeed, such data citizens might find that their "rights" to data are restricted if they attempt to access and use their data or the data of others in these ecologies. Data citizens, in this sense, are not necessarily working in a deliberative or democratic vein. Instead, they are activated through participation that does not lead to a "right to" anything as such. Here, participation could become the basis for further de-democratization, even while the term "data citizen" is mobilized to suggest otherwise. One of the more sinister uses of the term "citizen" is analyzed in Ruha Benjamin's discussion of the mobile app Citizen, which allows users to undertake community surveillance in their neighborhoods. Such practices typically exacerbate racial profiling, where the "norms" of social discrimination can become encoded in the use of this "citizen" technology to monitor urban activity.[15] At the same time, the makers of this citizen-oriented app present it as contributing to the "democratization of the 911 call," where assistance is only a watchful neighbor away.[16] Here, a citizen is less a democratically engaged subject and more a surveillant node reinforcing inequalities.

Many studies on data citizens focus on technologies and data generated through social media or wearables, through which particular formations of surveillant operators or consumer–subjects materialize.[17] While citizen-sensing technologies could as easily reinforce these modalities of consumer–subjects, there are other ways of engaging with the possibilities of data as they support data activism and generate counter-data actions of resistance and self-determination by contesting the "truth" of prevailing forms of data.[18] Data can facilitate social organizing and advocacy.[19] Such practices can generate different subjectivities and affective engagements with the conditions observed and acted upon. They also form power dynamics and spark calls for data justice.[20]

As discussed throughout this study, citizen-sensing technologies are meant to enable particular forms of data citizenship by encouraging involvement in environmental problems. Plugging in, activating a digital toolkit, and joining a disparate community of users: these are seemingly the steps to follow to mobilize the right to clean air. Yet the processes of sensing environments, collecting data, documenting, and addressing environmental harm do not typically lead to such well-equipped political subjects. While considerable work can go into collecting and analyzing data sets, citizen data can easily be overlooked and ignored. Rather than unfold a frictionless form of engagement, citizen-sensing toolkits and the citizen data they generate can lead to even more complex struggles with urban environmental life. Data citizens are, then, figures of struggle.

Indeed, even the right of citizens to monitor environments can be thrown into question, with practices, protocols, and devices subject to legal intervention and scrutiny. The right to monitor environments is not guaranteed, and in some countries the practice has been deemed illegal. In the United States in 2015, the state of Wyoming attempted to outlaw many forms of citizen monitoring, including photography, after concerned citizens documented *E. coli* in water samples from streams, where the source of pollution was from grazing cattle.[21] The bill sought to forbid the collection of "resource data," including data from air, water, soil, and vegetation, by designating this activity as trespassing, even if occurring on public land. However, the Tenth US Circuit Court of Appeals sent the case back to the lower courts, where the pending law was thrown out on the basis that it violated the right to free speech, which includes the right to petition the government.[22] While the right to monitor in this case was upheld, many state-level ag-gag laws in the United States still prevent documentation of meatpacking plants, for instance.

Yet the right to monitor instantiates more than a right to speech. It also instantiates a right to participate and a right to environments.[23] Such rights often do not feature in conceptions of citizenship that are based on a detached if deliberative subject. Data citizens form through evidentiary practices that document worlds of experience. In this sense, data citizens do not materialize as processors or objects of data. Instead, they form through struggles over the right to data and the right to environments that such practices would activate. Citizen sensing and the data it generates can document individual and collective grievances about pollution, development, displacement, and dispossession. Data citizens are not identifiable here through the usual membership categories of nation-state or consumer technology. Instead, they form as particular political subjects, relations, and collectives by working with and through data. Such data practices co-constitute distinct political subjects and worlds that would be sensed.

Data citizens are as likely to materialize through struggles with the erosion or absence of rights as through the inability or futility of appealing to rights. Evidentiary techniques become a process for materializing data citizenship. Such techniques can transform in urban worlds, especially when rights fail to materialize. Citizen data that document urban change and conflict can rework both data citizens and processes of urbanization. It is at these sites of struggle that multiple other forms of data citizens proliferate, less as fully formed political actors and more as persons and milieus attached to, yet haunted by, the promises of democratic life.[24] Data citizens materialize in this way as another version of citizens of worlds that, in resonance with the atmospheric, instrumental, and speculative citizens discussed in previous chapters, require exchanges with milieus to come into formation. In other words, data citizens are distinct expressions of citizens of worlds. These citizenships form not just by gathering and circulating data but also by mobilizing data to make more breathable worlds.

Rights, Citizens, and Worlds in the Making

Rights often manifest in digital and social-media technologies as the right to privacy, the right to be forgotten, the right to data protection, and the right to open data. However, this discussion proposes another way of thinking about how rights to breathable worlds, along with multiple other rights, materialize through citizen data practices. The use of environmental monitoring technologies can activate different rights in the making. This process of remaking and creating rights changes the relation and constitution of the "citizen" in "data citizen." Data can become a way to track, document, and concretize lived urban experiences. Rights can also encompass relations that signal distinct ways of being in and for worlds. Such an approach expands rights beyond a discursive claim[25] to constitute rights as spatial-material practices and formative relations for making and sustaining worlds. The right to relations, the right to collective life, and even the right to responsibility might materialize in these recast ways of forming rights.[26] The power relations that inform the becoming of citizens are not just exchanges with those who would govern.[27] They are also shaped through exchanges with more-than-human entities and environments, where power is situated, lived, and potentially transformed.[28]

If rights can be characterized as more than discursive claims, then they might be differently approached as relations, dispositions, orientations, infrastructures, collective feelings, atmospheres, and distributed practices that encompass more than an individual rights-bearing citizen. Here, political subjects form by tuning in to and activating environments and environmental problems. These are citizens of worlds. Citizens could form through the conjoining of multiple entities that make possible the conditions of political subjects, as in the case of citizen sensors (where sensors could be technical or organismal in form, as discussed in the next chapter). The citizen–subject materializes through relations with worlds: this is a condition of sense-ability and breathe-ability. Data citizens express a right to data and a right to worlds. Different possibilities for being and becoming citizens of worlds are constituted through these exchanges with worlds. To be and become citizens of worlds requires the development of practices and relations that are in constructive and formative exchange with those worlds.

Citizens, rights, and worlds are all together in the making. Practices that express a right to make breathable worlds remake environments and inhabitants. If data practices contribute to the formation of citizens as political subjects, then they are also fused with the articulation of rights and worlds in the making. This is one way of articulating what Étienne Balibar has referred to as the "continued invention of democracy" that unfolds through struggle and the pursuit of justice.[29] Such invention necessarily extends to citizens and worlds in the making. Sensor-based data practices constitute and activate data citizens through engagement with data and devices and through the struggles that data support and mobilize.

Evidentiary practices create and operationalize citizenship not only as an articulation of preexisting rights to be upheld but also as the ongoing formation of social, political, and environmental struggles. The pull toward rights not yet realized can shift the usual way of designating and engaging with problems. Rather than operating as a guarantee of an abstract and stable condition of citizenship, citizen data that document air pollution make evident how rights materialize as *prospective* practices or as sought-after relations.[30] The right to clean air indicates how to work toward transformed and more equitable collective atmospheres as worlds in the making, and how to become citizens of worlds.[31] "Citizens of worlds" is a concept that signals these prospective practices of political engagement, where the formation and exchange of subjects and worlds are a central part of what constitutes the sense-ability and breathe-ability of sociopolitical life.[32]

The Right-To as How-To

The *right-to* gives rise to multiple practices of the *how-to*. As discussed throughout this study, how-to consists not simply of following instructions but also involves developing practices that engage with the multiple struggles, techniques, and strategies that unfold through working with data and attempting to sustain, create, and transform urban worlds. The *right-to* proposes *how-to*, including how to be in exchange, how to generate environmental actions, and how to sustain political engagements. In this way, the imperative mood of the how-to invokes collective responsibility rather than a command for how to undertake such projects. The right-to is a form of how-to that works toward more democratic conditions. Distinct modes of citizens and citizenship form through pursuing the right to as a practice of how-to.

Within this context, sensor-based data practices can propose rights that become instruments for making more breathable worlds. Citizen data can generate open-air instrumentalisms, where rights are claimed, instantiated, circumvented, and reworked as part of the conditions of more livable and just environments. Such reworkings occur through practices that generate different forms of data, implement or challenge the observational techniques and infrastructures of experts, and make alternative proposals for urban environments. The right to data and the right to breathable worlds contribute to tools and toolkits that seek to make openings, lead struggles, and work through practical situations in and through which urban projects form.

Rights are another sort of instrument that contributes to the open-air instrumentalism of this shape-shifting toolkit. Yet while the right to data can co-constitute the right to make breathable worlds, such rights can also be difficult to realize for those who are pushed to the edges of urban life. In the context of citizen-sensing practices, the right to data and the right to breathable worlds are not established political or legal conditions that would serve as simple levers for fixing polluted environments. Such rights indicate, but do not guarantee, additional ways of working toward more livable environments. Instead, they are part of a broader open-air toolkit that seeks out strategies to cultivate more breathable worlds.

The right to breathable worlds raises the question of praxis, of how to engage in different configurations of theory and action. Citizenship is a sited, collective, and relational practice that activates environments in different ways. "Citizenship is the practical site of a theoretical existence," as Lauren Berlant notes.[33] The practical sites of citizenship involve the active forming, testing, challenging, undoing, and remaking of political engagements and political subjects. This research on citizen sensing forms a collective inquiry into the conditions of practical engagement that materialize along with experiments in different urban inhabitations.

These practices further demonstrate commitments to struggle for worlds that might be more livable, but they are unevenly available. Within these struggles, failure is likely. Failure, however, is not the flip side of success but rather a recognition of the pitfalls in praxis, where struggle can encounter the "impasses of the political."[34] In these moments of impasse, the reinvention of citizenship, rights, communities, and the worlds that are made and sustained can appear more viable. As Berlant writes, "It may be a relation of cruel optimism, when, despite an awareness that the normative political sphere appears as a shrunken, broken, or distant place of activity among elites, members of the body politic return periodically to its recommitment ceremony and scenes." Such recommitment can involve paying attention to how political formations hold together, how they fall apart, and how they might be remade toward a "more livable and intimate sociality."[35] Rather than bundling rights into a practice available to a universal if diverse grouping of citizens, such an approach might instead tune in to the plurality of political subjects and the struggles they encounter when attempting to invent, articulate, materialize, or transform rights. In this way, struggle becomes the basis for realizing even more—and expanded—modes of citizenship.[36]

The open-air aspect of this investigation necessarily involves questioning rights-as-instruments to consider how different approaches to breathable worlds materialize or are thwarted. Although various rights might be claimed through the practices of data citizens, there are many ways in which rights do not generate more democratic environmental engagements. Practices of citizen sensing of air quality in South East London do not so readily realize the rights they pursue. However, they do potentially reinvent rights and modes of citizenship through alternative political engagements. Such practices are often less utopian or triumphant. They turn up at the frayed edges of citizenships that are denied or never realized, often because of inequalities that include but are not limited to conditions of gender, race, or economic status.[37]

Data citizens might in this way become less oriented toward the overt ambitions of rights and more engaged with finding provisional techniques for staving off and surviving dispossession, pollution, and injustice that often accompany increasing urbanization. A right to make breathable worlds and a right to data offer powers of engagement and transformation that can seem out of reach for many urban dwellers. Such rights in the making could promise democratic participation that is difficult, if not impossible, to realize.

Citizen data can at once displace and reinvent rights, especially as they fail to address environmental problems. Rather than claiming rights, citizens could mobilize data as a persuasive tool for making arguments in support of urban life. When, for instance, an appeal to the right to housing seems too complex or politically impossible to undertake, urban inhabitants might instead demonstrate how new construction is not affordable to local residents. Data-based observations and arguments about unlivable urban conditions become a stopgap measure to sustain urban ways of life that are continually under threat but for which rights are often not enforceable or do not exist. Some researchers suggest that rights are a way to guarantee environmental protection in a way that citizen data cannot, since citizen data can be readily challenged as inexpert and imperfect.[38] However, environmental rights are often difficult to enforce and uphold, even when supported by the most "expert" forms of evidence. Indeed, the perceived ineffectiveness and unevenness of rights could mobilize data collection. In other words, if rights were effective, then people would not necessarily be so inclined to undertake environmental monitoring, since presumably states and other institutions would perform these functions to uphold environmental rights. While an abstract designation of rights might promise an ideal condition, it is often through more contingent practices such as citizen data that rights differently mobilize as subjects and worlds in the making.

Data citizens might be most likely to materialize in situations when the right to clean air becomes difficult to sustain and where rights fail to support struggles for more breathable worlds. People who may not feel that rights are a clear point of political attachment create evidentiary practices to challenge the dispossession, environmental damage, and injustice of neoliberal urbanization. Citizen data could, in this sense, be a practice that manifests where rights break down or are not yet established.

Data for Black Lives is an example of such a movement that involves developing alternative data-collection and data-analysis techniques to create new narratives about Black people's lives while also demonstrating how systemic racism attempts to maintain inequality.[39] As these practices demonstrate, rights are not always self-evident, since there are many rights that Black people have that are often not protected or observed. Many data-oriented arguments could be made that do not clearly reference rights. Instead of data configured to support "universal" rights, data could instead be mobilized to support struggles for everyday survival and dignity in the absence or partial enactment of rights. This is what Data for Black Lives founder and executive director Yeshimabeit Milner refers to, in the spirit of W. E. B. Du Bois, as a way to rework data practices away from the destructive uses to which they have been put to reinforce and propagate racism. By creating new data practices, including analysis and visualization, Milner suggests that other possible ways of evidencing Black people's lives might be sparked.[40] These practices of computing otherwise could activate protest, accountability, and collective action while forming different narratives and rights.[41]

By documenting air pollution, people come up against the inertia and failures of politics. In multiple and diverse struggles to engage in urban democratic processes, the right to data becomes one way to express a right to breathable worlds. Yet these pursuits can also be derailed through sclerotic urban governance structures, rigid formations of expertise, or exclusionary processes of urban development. Such data practices can then constitute and propose ways of being data citizens. By undertaking environmental monitoring, citizens mobilize rights to data, air, breathable worlds, and political life. These practices observe, document, and remake urban life. They propose conditions for being and becoming citizens of worlds. As propositions, they unfold as open-air instrumentalisms and how-to practices. In this way, such practices are guides for working toward more breathable worlds that have been tested, implemented, and that are still in the making. In the next section, I discuss how the Citizen Sense research group worked with communities in South East London to sense the air and to undertake practices for pursuing the right to breathable worlds.

CITIZEN DATA IN PRACTICE

Following the research focused on fracking and pollution described in the previous chapter, this second phase of Citizen Sense research studied citizen sensing of air pollution in urban environments. During nearly two years, from spring 2016 to late autumn 2017, we collaborated with residents of the neighboring wards of Deptford and New Cross in South East London to monitor air quality in relation to traffic, development, and industrial emissions. These neighborhoods are sites of former industry—dockyards and a historic naval shipyard—as well as community markets, housing estates, and an incinerator. An area that has been marked by economic deprivation and inequality, unemployment and limited job opportunities, Deptford and New Cross also have larger Black and minority ethnic populations than many other parts of London.[42] The area has been the location for ongoing struggles over environmental injustice, including the siting of the incinerator in New Cross in the 1990s that continues to operate today.[43] As has been well established in the UK, air pollution tends to affect people living in lower-income areas, and incinerators are also far more likely to be sited in deprived areas.[44]

However, the area has an even longer history of its residents engaging with the problem of urban air quality. While living in Deptford in 1661, John Evelyn wrote one of the first texts on air pollution in London, *Fumifugium*, a text that some residents and community groups continue to reference when making a case for mitigating air pollution and improving the urban realm.[45] With a rich history of organizing for social justice, communities in Deptford and New Cross have undertaken projects to respond to, or intervene within, processes of development and the problem of environmental pollution.[46] A 1999 study, *Surviving Regeneration*, documents the looming threat of increased development in the Deptford area and proposes how to mitigate the effects of environmental damage. "For

Figure 3.1. Low-emission-zone boundary sign and traffic camera on the Old Kent Road; traffic in South East London. Photographs by Citizen Sense.

some time," the study notes, "South East London has been characterised as 'the soft underbelly of the capital,' a place of industrial dereliction, cheap sites and demoralised labour."[47] The text documents how, in this area of "tides, wildlife, dereliction, rubbish, hope," numerous surveys were undertaken to attempt to guide regeneration toward less socially and environmentally destructive outcomes.[48] These surveys incorporated assessments and environmental monitoring of the area, including rubbish in the creek; archaeology and history; the biodiversity of birds, mammals, vegetation, fish, and invertebrates; the toxicity of creek water and mud; flood defenses; and community heritage. The text also documents how engagement with local people was an uneven process, often thwarted by the relative absence of policy makers at community meetings.

Twenty years later, Deptford and New Cross continue to experience waves of development and densification that contribute to significant changes in the urban environment, along with struggles related to unequal participation in political processes. The urban fabric in this location continues to be reworked and gentrified through new development schemes, master plans, and public–private initiatives.[49] Meanwhile, the increase in traffic and housing in this area and throughout London has led to further congestion and air pollution.

Within this context, and seeking to learn more about the environmental monitoring practices already being undertaken in these two wards, we researched and contacted community groups to learn about local initiatives that sought to address urban environmental problems. Through multiple projects and campaigns, residents were engaged in monitoring air quality, counting traffic, assessing the state of urban trees, and documenting disruption to green spaces and biodiversity. In the process of learning more about the projects and campaigns underway, we met with people caring and advocating for parks, high streets, and housing estates, as well as those campaigning for better transport conditions across the wards. In our conversations, residents brought up environmental changes they had experienced and told us about environmental monitoring they had organized to contest road use and to monitor dust pollution across construction sites. Concerns about air quality were intertwined with wider urban environmental problems related to the rapid pace of changing land use within the area, primarily through the development of high-density, high-end housing.

These practices came together to support cases for improving the urban realm, which were made through local meetings, planning applications, and campaigns. For one citizen-monitoring project, "Don't Dump on Deptford's Heart," residents installed diffusion tubes for monitoring nitrogen dioxide in order to contest the proposed development of a Nationally Significant Infrastructure Project, the Thames Tideway Super Sewer, constructed to update the decrepit

Figure 3.2. Dustbox monitor and installation at participant location in New Cross Gate; *Deptford Is Changing* text documenting study of the urban realm. Photographs by Citizen Sense.

Figure 3.3. Signs documenting local community organizations and protest against new development. Photograph by Citizen Sense.

Figure 3.4. Deptford Lounge and Library, where Citizen Sense held a monitoring workshop and made Dustboxes available for loan. Photograph by Citizen Sense.

London sewage infrastructure that contributes frequent wastewater discharges to the Thames.[50] With the data they collected with these analog monitoring devices they were able to document the poor state of air quality throughout the proposed development area and surrounding context.[51] Despite these efforts, the Super Sewer was approved for development, and construction began on the project. While the Super Sewer is meant to mitigate the problem of water pollution, especially in the River Thames, residents feared it would exacerbate the problem of air pollution by producing emissions both during construction and during operation of the sewer at pumping stations and ventilation shafts.

Indeed, multiple development sites were and continue to be actively contested by residents. One small area, Creekside, located on the eastern edge of Deptford, had at least five separate development sites under construction during this monitoring study. Residents suspected that such developments were likely contributors to increased air pollutants throughout the development life cycle. From demolition and site clearance to construction and heavy-goods vehicles, as well as increased density and traffic once development is completed, the environmental effects of construction can be felt for years. At the same time, the impacts of construction are inevitably bound up with the relative economic and social injustice related to new developments as people are displaced from rented and public housing and often not able to afford to live in the area once the brunt of negative environmental effects from development has been endured.

In order to contest development, as well as to seek compensation from developers in the form of community development funds, many residents and community groups had undertaken environmental monitoring projects to demonstrate the ill effects of living with constant construction. From traffic counts to air-quality studies using diffusion tubes, local citizens generated multiple forms of data about their environments. They also encountered, analyzed, and used data from governmental entities and industry, including in the form of planning documents in online portals; community meeting minutes; environmental impact and environmental assessment reports; official air-quality data; construction-company self-reporting on pollutant levels (including air, noise, and light); utility-company data on pollutants from national infrastructure projects (including air and noise); tree-map data designating tree locations and numbers; tree-removal applications; social statistics on population, density, and income; social-media data (including Twitter and Facebook); crowdfunding data; petition data; word-of-mouth data (often about proposed development schemes); and many more types of data on the London Data Store and the Lewisham Borough website.

In these numerous engagements with environments, data, and governance, people became data citizens in part through wrestling with multiple forms of data and attempting to articulate a right to clean air, a right to participate, a right to the environment, and a right to make breathable worlds. Data became a means to express and materialize rights or create rights in the making. Citizens analyzed publicly available data, sought data through FOI requests, documented events and environmental disturbances by creating their own data sets, and communicated and contested changes to the urban environment through these multiple data sources. They also produced their own data to counter or qualify government statements and industry claims. They did so in the absence of official monitoring networks or where data were not sufficiently analyzed or acted upon, often because of austerity measures.

These multiple data practices constituted data citizenships by creating new citizen data and by linking different data sources to create particular accounts of urbanization that could intervene in these processes. These practices attempt to materialize rights—both as claims and prospective lived conditions. They present evidence even when appeals to rights are not heard or realized. In this context, the Citizen Sense research group collaborated with residents to develop a citizen-led air-quality monitoring network to research how data citizenships might materialize or transform by generating and integrating data into these multiple data practices.

How to Mobilize Citizen-Sensing Infrastructures

Along with learning more about community concerns and campaigns in the area, we worked with residents to develop a toolkit that could grow into a citizen-sensing infrastructure for monitoring air quality.[52] Our collaboration with community groups, residents, and workers involved learning more about their diverse data practices, whether in the form of environmental monitoring or analyzing government data sets, while also engaging in meetings, workshops, walks, and site visits to explore the particular uses of citizen-sensing technologies in this part of South East London. Far from acting as experts with a singular way of accounting for urban environments, we contributed as co-researchers to data practices that joined up with existing community initiatives.[53] We were, in the process, also becoming data citizens as we collaborated with inhabitants and learned more about their concerns for and ways of documenting the area.

For this second phase of research, we were in part drawing on our previous work on sensors and air quality developed in relation to fracking in rural Pennsylvania (described in chapter 2). Yet we were also responding to the area by developing sensors specific to this urban location. Rather than use an off-the-shelf device such as the Speck, we built a new prototype device, the Dustbox, which monitored $PM_{2.5}$. As previously discussed, $PM_{2.5}$ is a particularly hazardous air

Figure 3.5. Installing and repairing Dustbox monitors in Pepys, South East London. Photographs by Citizen Sense.

Figure 3.6. Setting up a Dustbox at Besson Street Gardens in New Cross Gate. The lowest levels of air pollution were documented in this highly planted and garden-based monitoring location. Photographs by Citizen Sense.

pollutant made up of a range of different materials that can lead to cardiovascular, respiratory, and neurological diseases, among other conditions that are in the process of being studied and documented.[54] However, there was a relative absence of regulatory infrastructure for monitoring particulate matter in Deptford and New Cross, and so the development of a citizen-sensing network offered the possibility to better understand concentrations and potential sources of this pollutant.

We created the Dustboxes based on the form of pollen and contaminated soil particles when magnified under an electron microscope. Fabricated through a 3D-printing process and cast in black ceramic, these small plug-and-play monitors used the widely available Shinyei PPD42NJ particle sensor unit, installed in numerous low-cost and DIY monitors in circulation at the time. The Shinyei particle sensor applies heat and an infrared light scattering technique to circulate and sense particles with a diameter of 1 μm or larger.[55] A receptor receives the scattered light from the particles to measure the relative opacity of air entering the sensor chamber, which is transformed into a pulse signal that can be further converted into particle concentration. The Dustbox monitor also included a custom-printed circuit board, an Electric Imp Wi-Fi module, and a fan for circulating air. We developed the Dustbox as an affective and tactile device that would resonate with the often gritty environmental conditions of this area in South East London while also circulating as an engaging citizen-sensing infrastructure.

Along with investigating the citizenships that might be activated or mobilized through setting up a network of Dustbox particle sensors, we were interested to understand how the Dustbox could operate in an urban setting where there was a well-known problem with air quality but not necessarily a single emissions source that could be readily identified. Yet air pollution was just one of many urban problems that people sought to address. In contrast to visions of the smart city that imagine the urban setting as a blank canvas for implementing digital designs and wiring up citizens, these were spaces where citizen sensing and sensors operate among an already sedimented and established set of processes and concerns. Our project sought to investigate how sensors and data practices could establish the relative intensity of pollution in the area, while proposing different ways of activating rights and citizenships in relation to ongoing urban struggles.

As part of the collaborative development of the Dustbox as a citizen-run air-quality monitoring infrastructure, we worked not only with community members but also an array of collaborators, including atmospheric scientists, so that we could calibrate the Dustbox in relation to the "official" air-quality network in London. This process involved co-locating the Dustbox with regulatory-standard

Figure 3.7. Dustbox Logbook with monitoring instructions and space for recording observations; Citizen Sense workshop for mapping where Dustboxes could be set up in the Deptford and New Cross areas. Photographs by Citizen Sense.

instruments at the Marylebone Observatory run by the London Air Quality Network and the UK's Department of Environment, Food, and Rural Affairs (DEFRA).[56] We compared the relative particulate-matter measurements across Dustboxes, as well as comparing levels with measurements from regulatory devices. This colocation and calibration process allowed us to make a notional conversion of raw particle counts (as a measure of voltage) into micrograms per cubic meter, a unit of measurement that would allow participants to make indicative comparisons between the Dustbox and the official monitoring infrastructure.[57] While citizen data express multiple registers of urban experience, the calibration techniques also created one register for comparing data across different monitoring sites and instruments.

As the Dustboxes were assembled into provisionally workable sensors, we built on methods developed through our fracking-based research to organize a public workshop and walk in late October 2016. The workshop brought together residents, community groups, health researchers, and an assembly member of the Greater London Authority to discuss air quality in relation to the changing urban environments of South East London. During the workshop, we described the Dustbox and related tools for analyzing data, discussed previous monitoring efforts in the area, and identified additional monitoring sites based on community knowledge of emissions sources. Participants mapped locations they intended to monitor as they noted likely pollution hot spots and sites of changing land use. We also considered how different observations of air quality might be recorded, since sensors could provide a more real-time and quantitative way to tune in to air quality, while recorded observations of sound, smell, construction activity, traffic, and other urban events could provide parallel ways to configure data as evidence. We provided a Dustbox Logbook where such observations could be noted, which as parallel forms of data would later inform the analysis of citizen data and composition of data stories.

During this introductory workshop, we also set out on a walk to look at key sites of construction activity, roadways clogged with traffic, and industrial sites so that we could discuss where to monitor and how to study changing land uses.[58] As part of the walk, we looked at existing sensors and monitoring infrastructure installed in the area, including air quality and noise monitoring underway as part of the Super Sewer development in Deptford. We discussed whether we could access the data from these monitoring sites, since the Thames Tideway data were not readily accessible to the public, even though this was a national infrastructure project. Although environmental monitoring was taking place in multiple forms in this neighborhood, the data were rarely open to wider use. In the absence of such data from small developments to large infrastructure projects

Figure 3.8. Walk in Deptford to investigate proposed monitoring locations and possible sources of air pollution. Photograph by Citizen Sense.

Figure 3.9. Air-pollution monitoring infrastructure sensing wind speed and direction with air pollutants in Deptford to document possible emissions from the Thames Tideway Super Sewer construction. Photograph by Citizen Sense.

across London it could be difficult to gauge whether local pollution events might be occurring. Participants considered how they might request data at meetings with the Tideway organization. They also assessed which other air-quality data might be available nearby, since at the time the borough of Lewisham had just three air-quality monitors installed over a large area, which would provide only a rough estimate of air pollution at the actual sites of major construction projects.[59]

After walking around these development sites, we moved to a local pub to distribute Dustboxes and talk through how to use the devices and where residents might monitor. Many participants borrowed Dustboxes during this workshop event, while others checked monitors out of the local library or contacted us directly to pick up a monitor. The Dustbox Logbook included setup instructions so that participants could install the devices themselves, although we ordinarily arranged visits to monitoring sites to help participants install and troubleshoot their devices.

The Dustbox infrastructure grew into a changeable and fluctuating infrastructure. As new people began monitoring, others paused or stopped monitoring. We set up a citizen-sensing network that included up to thirty Dustbox sensors monitoring $PM_{2.5}$. However, the number of Dustboxes running varied throughout the monitoring period spanning nearly ten months from December 2016 to September 2017, with consistent monitoring at eighteen locations over seven months. We made numerous visits to monitoring sites to install devices, connect them to Wi-Fi networks, find suitable outdoor space for monitoring urban air, and make adjustments along the way as devices went offline or required repairs.[60]

The loaning of devices was just the beginning of a more extensive process of setting up Dustboxes, since each monitoring spot had its unique requirements and idiosyncrasies, from unusual Wi-Fi router configurations to complex siting arrangements. Dilemmas arose about where to place the Dustboxes so that they would not become soaked in rainstorms or be nicked by passersby but would also be located at suitable heights for detecting pollution.[61] Technical investigations ensued to seek the best arrangement for the monitors in what were often collective outdoor spaces. Dustboxes became part of the furniture in instances of home placements and were also stealthy yard and patio ornaments lurking under out-of-use barbeques and other garden architecture. And the logbook took up residence along with other everyday items, marking and expressing commitments to working toward collective practices for building breathable worlds.

The collective process of setting up a citizen-sensing infrastructure became a way to materialize rights to data, air, and breathable worlds. These practices of building and connecting monitoring infrastructure express a right to constitute

Figure 3.10. Dustbox installation and setup in the Creekside area of Deptford. Photographs by Citizen Sense.

and transform modes of citizenship through attention to and engagement with urban milieus. Rights in this sense are expressed less through legal challenges and more through collective inquiries into the "state of the air," where the "state" is not a nation but an ongoing sociopolitical project of atmospheric citizenship as it meets urban, environmental, and data citizenships. While data citizens in part materialize through sensor-based data practices and infrastructures, they also mobilize along with the activation of rights that might help mitigate and address air pollution.

How to Figure Citizen Data

Inevitably, the question arises about what can be done with data from these sensor devices, especially given the considerable effort involved in setting up monitors. How might it be possible to figure citizen data into forms of evidence that can support and mobilize rights to breathable worlds? Citizen data do not merely replicate or challenge official data sets. Instead, citizen data can figure different worlds and call them into being by expressing lived experiences, recasting approaches to air pollution, and proposing different configurations of urban environments.[62] I build on Donna Haraway's discussion of figuring to consider how this is a way of configuring, numbering, narrating—as well as creaturing—data.[63] "Creaturing" is a concept that I previously developed to express how data obtain relevance through the distinct worlds in which they are generated and have effect.[64] Data in this sense are not simply descriptive of worlds. Instead, data and worlds are co-constituted as distinct modes of inhabitation and conditions where data have relevance. Citizen-sensing practices creature and story air-pollution data by generating problems to which data can respond and attach and for which data come to have significance. Creaturing is a process whereby data can come to figure, or in other words, to matter. But as I suggest here, different creatures of air-pollution data can also create sites of struggle in terms of the pluralistic data and urban worlds that matter or are sustained, overlooked, or extinguished.[65]

In this investigation into how citizen data can contribute to the formation of citizens of worlds, we then considered how to build on and develop analysis techniques that might distinctly figure and creature citizen data by connecting that data to extended infrastructures and practices. Based on our earlier fracking research, we adapted our Airsift data-analysis platform so that Dustbox data could be viewed and analyzed in relative real time. Monitoring sites were mapped with fuzzy locations, and data were open and available for viewing, analyzing, plotting, graphing, and downloading. We pulled in data from the London Air Quality Network API (application programming interface) to compare citizen-sensing locations with nearby regulatory monitors. With this toolkit, people could investigate, review, and analyze their own data as well as other data in the net-work. In this sense, we developed the Airsift toolkit to enable DIY data analysis. This approach extended the attempt to democratize monitoring by testing ways to democratize data analysis, while keeping in mind the pitfalls of democratization as a techno-political process and promise.

Because the air-quality data were not necessarily easy to analyze for people new to atmospheric science, we collaborated with participants to host data work-shops and drop-in data tutorials to look more closely at patterns emerging in the data. In these meetings, we introduced the Airsift tool, worked through analyses of citizen data sets, compared data across different monitoring sites, and strat-egized about where else to place monitors and gather data in support of com-munity projects. These exchanges elicited questions about how to engage with pollution in ways that connected to experiences, while also developing techniques for analyzing data and making atmospheric science legible within broader forms of urban engagement. We found that this spatially dense network of citizen-sensing devices allowed us to zero in on particular urban patterns, processes, and distributions of pollutants. Often working at the scale of one-hour and twenty-four-hour mean levels of particulate matter, we could analyze the specific and comparative timing and distribution of pollutants in the area, which allowed us to gain a much more detailed picture of urban activities underway.

With these analysis techniques, we discussed how data could assemble into different forms of evidence that might be useful for informing policy, neigh-borhood plans, or other initiatives that responded to development, construction sites, and transport in this rapidly changing part of London. Using our Airsift toolkit we plotted times of day and week when pollution was occurring. We often found increased pollution toward the end of the week, with a decrease on Sun-days (no doubt related to traffic, the primary source of pollution in London). Events such as Bonfire Night become clearly visible as elevated episodes in the data due to fireworks. And shared pollution patterns were spotted across local and regional sensors, depending upon pollution sources.

As a register of urban environmental processes, the Dustbox data began to unfold in relation to everyday urban life. Moments when air pollutants regis-tered at particularly high levels became an event where we would pool collective knowledge about industry activity, fires, pollution drifting in from Europe, or intensive construction might help to explain peak readings. We also worked together to collect and analyze data from the London Air Quality Network (using air-pollution data and alerts), from Lewisham Council (in the form of planning documents and air-pollution apps), the UK Environment Agency (to document industrial monitoring sites), and the Greater London Authority (to incorporate tree maps and other data). In this sense, quantitative sensor data did not provide an absolute or definitive figuring of urban events. Instead, citizen data featured most significantly when multiple observations and other forms of data came together to corroborate and transform lived urban experience.

If people collect data but those data are closed down or inaccessible to analysis, then this practice might more accurately be referred to as crowdsourcing, since the data are owned and mined by actors other than the citizens who collect the data. Here, participants generated their own data that were open for further use, including through Airsift as a DIY data-analysis toolkit. But data were more than "open" in the usual sense, since they were not simply a .csv file made available by a government entity in a data repository. Instead, the data were embedded in situated monitoring and data-collection practices, as well as available for open analysis and mobilized within projects to advocate for the urban environment. In this sense, data are less an enumeration of individual behaviors or conditions and more a collective resource and infrastructure that can support exchanges across citizens and worlds. Such practices express a breathability of data as much as a breathability of worlds.

Based on the multiple meetings, workshops, and conversations with participants and residents, we collated our collective findings from the ten months of Dustbox monitoring in seven online and print-format *Deptford Data Stories*.[66] We crafted the data stories as a collaborative method for figuring citizen sensor data in the form of numerical measurements, maps, on-the-ground observations, images, and narratives about activity in the urban environment. The data stories composed the citizen data into distinct accounts of air pollution that could narrate overlooked urban experiences while enabling collective proposals for transforming environments toward greater livability.[67]

In analyzing the citizen data, we found that major traffic intersections and construction activity, as well as the Thames, all showed up as likely pollution sources, often at levels well above the WHO twenty-four-hour guideline of 25 µg/m³ for $PM_{2.5}$. We also found that green spaces and sheltered gardens often had much lower levels of $PM_{2.5}$. The process of arriving at these findings involved discussions about urban activity and likely emissions sources, queries about distinctive patterns in the data, site visits to inspect pollution activity, and negotiations about how and when data might be more widely circulated so that conversations could be held with local government and community groups.

Processes of collecting data generated ways to figure, creature, and materialize rights in the making, including the right to data, the right to clean air, and the right to make breathable worlds. And yet these rights were unevenly acknowledged by local and national government, industry and developers, and other "stakeholders." While such rights are often not enforced or even recognized, data can aid attempts to counteract the failure of rights or to activate the possibility of such rights in the making. The next section considers more specifically how the right to breathable worlds—as a prospective right—materialized through data stories and community projects. Within this context, data practices differently addressed social, political, and environmental struggles by attempting to reinvent rights.

THE RIGHT TO BREATHABLE WORLDS

As we collaboratively analyzed citizen data and developed data stories, these citizen-sensing practices folded into ongoing community projects to defend and transform urban environments. Here, data citizenships materialized through practices that expressed the right to collect and analyze data and the right to advance proposals and implement projects for transforming urban environments as an expression of the right to breathable worlds. The data accumulated from multiple Dustboxes in South East London began to inform the co-constitution of citizens, rights, technologies, and material conditions. Before we publicized the findings from the citizen data, we hosted a workshop to review the draft data stories. In this event, we worked with citizens to review initial findings, make sense of data patterns, and compare plots and graphs with observations and experiences. As a key part of the workshop, we coauthored actions and proposals to address and mitigate air pollution in the area. Spanning from proposals for transportation experiments to the development of green infrastructure, air-quality monitoring campaigns, and construction controls, the actions responded to air pollution patterns by outlining concrete measures that connected to and supported ongoing projects and campaigns. The actions also formed a wish list for additional work that could be done to improve conditions of social and environmental justice. Here, citizen sensing joined up with citizen design, where democratized environmental evidence generated proposals to shape urban environments.

We published the completed *Deptford Data Stories* online in November 2017 and circulated a press release to local councilors, policy makers, the press, and other air-pollution researchers. London newspapers took up the findings, including the *Evening Standard*, which led their story with citizen data findings that pollution levels were more than six times the WHO's twenty-four-hour guideline

Figure 3.11. Dustbox installation near the Old Tidemill Wildlife Garden in Deptford; community-planning guidelines for establishing a neighborhood plan. Photographs by Citizen Sense.

for $PM_{2.5}$.[68] While the newspapers focused on moments when pollution was exceptionally high, the data stories emphasized the spatial and temporal patterns of pollution and how these could inform and support actions to improve air quality in a sustained way.

Nevertheless, the news about excessively high pollution levels compelled the local Labour MP, Vicky Foxcroft, to take up the citizen data findings and bring them to the House of Commons for a debate. She put her concern and question to the Leader of the House, Conservative MP Andrea Leadsom: "Research carried out by the Citizen Sense project at Goldsmith's [sic] in my constituency shows that pollution in south-east London reached six times the World Health Organisation limit on several occasions during the past year. Can we have a debate on this important public health issue?"[69] In response, Leadsom noted that "the Government are determined to tackle the problem of air pollution" and action was being taken to "encourage and help local authorities to pay for new pollution-free zones." At the same time, Leadsom noted that the Mayor of London should be "putting in place measures to reduce the poor air quality in our great city."[70] Here, citizen data circulated to the center of the UK government. While it presented a persuasive and even alarming record of air pollution, along with a set of proposals for how to address this problem, the evidence was met with relative platitudes when Foxcroft asked what action the Government would take.

For community groups, the local and national government's attention to air pollution in the area was a welcome development. At the same time, the findings and action points led to variable outcomes. Data here did not seamlessly unfold into action. Instead, the process of mobilizing data generated additional complexities. Far from the frictionless connections between data and action that some citizen-sensing devices promise, here citizen data became ensnared in ongoing struggles over urban environments, ways of life, and local governance. Yet the difficulties of taking action could, in another way, register as the very conditions that form citizens and citizenship. Democratic engagement requires possibilities for exchange, as part of what constitutes the breathability of political life. But these exchanges are also impeded, blocked, and shut down, even as people attempt to observe, contribute, listen, and be heard when communicating how their worlds matter. Action materializes through struggle as political subjects try to realize these formative exchanges. Struggle, however, is a condition that the promises of citizen-sensing technologies often gloss over when promising more streamlined democratic engagement.

Despite the news of elevated pollution levels, the data stories provided a way to figure citizen data as narratives and experience, rather than only present quantitative measurements. Citizen data did not seek to fulfill a regulatory function,

Figure 3.12. Air-pollution sources in Deptford and New Cross, including river vessels burning ship diesel; and traffic on the A2, a major South East London thoroughfare. Photographs by Citizen Sense.

Figure 3.13. Old Tidemill Wildlife Garden and community-generated architectural plans and proposals for the space. Photographs by Citizen Sense.

but asked different questions and provided alternative perceptions of air-quality pollution that connected to concrete proposals for action for urban environments and social justice. In this sense, citizen data practices did not merely demonstrate that air pollution was occurring or often exceeded regulatory guidelines. Instead, citizen data supported campaigns and projects for transforming the urban realm. These trajectories of data and action were mobilized to demonstrate attempts to address urban pollution and inequality, and to make more breathable worlds. The "findings" of the data stories became ways of figuring, creaturing, and proposing actions for worlds where this data might register and become relevant.

How to Pluralize Data Practices

Among the multiple sites where citizen sensing took place in Deptford and New Cross, we identified seven clusters of monitoring locations that became the basis for each of the seven data stories. One of these locations, Old Tidemill Wildlife Garden, was an area where many people were interested in monitoring, since they hoped to demonstrate that the green space was beneficial in mitigating particulate matter levels. Old Tidemill Wildlife Garden was originally a school garden that was turned over to the public when the school moved. The space became a wild green oasis within a heavily developed and polluted urban area, where community groups hosted forest schools and biodiversity workshops, organized community picnics and music festivals, built adventure playgrounds and tended vegetable gardens, and generally fostered the creative and activist energy for which Deptford is well known.

But in 2016, developers sought planning permission to build a range of market-rate and social housing in the place of the Old Tidemill Wildlife Garden. Peabody Housing Association developed plans to raze the garden and nearby block of council housing to build flats to address a housing shortage in the area.[71] Residents and workers were especially concerned about losing the community garden and adjacent social housing to high-rise (and more expensive) housing developments that would significantly alter the area. Here, the city was being made and remade, less as an expression of the right to build breathable worlds and more as a set of development projects that led to ongoing struggles over urban environments.

Residents, workers, and advocates for the Old Tidemill Wildlife Garden, including community groups Deptford Neighbourhood Action and Voice for Deptford, among many others, began a campaign to save the green space in response to what they felt were inflexible and unjust development plans. They worked with a design and architecture group to develop an alternative plan for the site that would preserve the green space and existing council housing while also allowing for new housing. The campaign and plans to save the space unfolded through a protracted struggle with the local council to draw attention to the significance of the green space, existing housing, and community ties built up over decades. People lobbied the local council, attended planning meetings, raised concerns at local ward assembly meetings, developed online campaigns on social media and websites, set up crowdfunding initiatives, worked with artists and designers to make films and host events, and publicized garden openings so that more people in the area would visit and learn about the space.[72]

In this context, several people who were engaged in the struggle against the development of the garden took up air-quality monitoring with Citizen Sense to develop yet another form of evidence that might aid their campaign. They sought to establish whether pollution was occurring on busy roads nearby and if lower levels of pollution could be detected within the garden area. As a result, we located monitors on balconies and outdoor spaces adjacent to the garden. Over a several-month monitoring period, the sensor data demonstrated a clear pattern of lower pollution in areas sheltered by the garden, and higher pollution in areas exposed to busy roadways on the perimeter of the garden. These findings spurred proposals, which the data stories included, to protect the garden as an important green space in the area and to augment and extend green infrastructure to address and mitigate pollution in the area.

Despite the soundness of the citizen data and the creative scope of the proposals put forward for how to preserve the garden while accommodating new development, the council remained unmoved by the findings or proposals. It voted to approve the development plans and to terminate its lease on the garden. While the council cited the need to address housing shortages as a rationale for developing the site, many dissenters noted that the development would not provide affordable social housing, yet it would remove access to a biodiverse green space. Here, citizen data did not facilitate or improve rights to data, environment, participation, or breathable worlds.

With the council's decision to forge ahead with the new housing and the turning over of the garden site to developers, multiple protests ensued. In August 2018, campaigners began to occupy the garden site, which included numerous mature trees, to attempt to halt its demolition. Protestors, as well as news reports of the garden occupation, frequently cited the findings from their citizen-sensing data, noting the problem of air pollution in the area and the role of green space in providing relief from pollution. Mitigation of air pollution became a key rationale for saving the garden, among other points related to protecting its biodiversity, preserving local housing and the community that had been established, and providing affordable housing.[73]

Figure 3.14. Save Reginald! Save Tidemill! campaigners photographed in the Old Tidemill Wildlife Garden in Deptford in October 2018. Photograph by Andy Worthington.

Donning gas masks and holding placards with phrases such as "Deptford Needs to Breathe!" and "Lewisham's Plans Cause Asthma," protestors voiced the need to address air pollution in the area and to make more breathable worlds not just by reducing pollution levels but also by safeguarding limited green space. People seeking to protect the garden made a film to persuade viewers of its unique characteristics while documenting how it made the area more breathable. The interviewee in the opening film sequence notes, "I just take a huge, deep breath is the first thing I do when I come in here," in reference to entering the garden and being immersed in other urban encounters and experiences that do not require sealing oneself off from harsh, traffic-clogged, and polluted environments.[74]

Despite these objections to the destruction of the garden, along with appeals to citizen data and multiple other forms of evidence mentioned throughout this chapter, the council persisted in turning over the site to developers by removing occupiers of the garden in October 2018. To do so, it hired the security firm

Figure 3.15. Protest installation at the demolition of Old Tidemill Wildlife Garden and nearby trees along Deptford Church Street and the Thames Tideway Super Sewer. Photographs by Citizen Sense.

County Enforcement, which had also participated in enforcement activities during the UK miners' strike in the 1980s. Many residents, workers, and protesters found this to be a particularly brutal and reprehensible measure taken by this Labour-dominated council.[75]

In February 2019 the garden was leveled and trees were demolished. Residents continued to monitor this destruction, using photography and video to document scenes of trees being ripped from the ground and cast aside as the site became a staging area for another London high-rise.[76] This was only one among many additional sites up for development in this compressed area. Shortly after developers leveled the garden and its trees, an additional seventy-four plane trees were felled to make way for the Super Sewer, and further trees were planned for demolition to build housing in which residents would not be able to open their windows during peak hours due to the elevated levels of air pollution in the area.[77] People continued to lodge ongoing objections to these relentless developments, appealing to their evidence gathered about air pollution. Councilors, however, did not review the citizen data and in some cases are purported to have boasted that they had made up their minds about the importance of housing development, irrespective of citizens' concerns about pollution and other damage to the environment and community fabric.[78]

When the local council failed to heed arguments about the unaffordability of the proposed urban housing—on the basis that there was neither a specific "right" to affordable housing, nor did people feel as though rights would be respected—citizens combined further data about air pollution to document the impacts from construction and traffic and loss of green space. Yet despite the established right to breathe clean air, as well as the right to participate in environmental decision making, these apparent guarantees of democratic engagement did not ensure that citizens would have a voice or be able to inform the shape and process of development in the area.[79]

The difficulty of mobilizing rights and the likely failure of attempts to realize rights can lead to the use of other tactics that attempt to contribute to the exchange, cultivation, and breathability of environments. Data citizens form through these practices of mobilizing evidence to support more democratic exchanges. Similar to the discussion in the previous chapter, at times the citizen data collection can offer alternative forms of evidence that enable exchanges with regulators and developers, politicians and the press. Data can document and generate different registers of experience, and enable possibilities to be and become citizens of worlds. The citizen-sensing practices and proposals narrated within the data stories developed in response to impasses experienced, and attempts to advocate for different approaches to urban environments.

Figure 3.16. Thames Tideway Super Sewer construction. Photograph by Citizen Sense.

Yet data can also produce their own disappointments, and rather than serve as a corrective to rights, they can compound problems of democratic unaccountability. Data do not always perform as expected. Data can be difficult to work with and analyze, but they can also lead to inertia and indifference on the part of regulators and policy makers who have fixed agendas and undemocratic practices. Some data count more than others, and data need to operate—and be creatured—within particular registers of relevance to be heard, apprehended, and mobilized. Data inequalities can take place not only in terms of whose data counts but also whose data can register as legitimate and significant.[80] These dynamics often unfold in relation to established dynamics of privilege and power, but they also are performed through more insidious dynamics of who gets to be counted as "the adult in the room." The "good citizen," as Claudia Rankine has suggested, is typically someone who does not speak truth to power, does not expose inequality, but does maintain a polite demeanor so as not to disrupt established conventions of civil and political conduct.[81] Such practices of the good citizen tend to reproduce rather than remake existing power structures.

Customary ways of exercising the *right to*, moreover, can assume a universal, privileged, normative, masculine, white, and actively enabled form of citizenship. Such citizenship practices would in part require that people struggle and confront injustices and exclusions, often in public forums and settings that favor some voices over others. In attempting to exercise these rights, many struggling urbanites could be exhausted by the crushing indifference of political processes.

As Berlant writes about such political engagements and attachments, people are "worn out by the promises that they have attached to in this world."[82]

Citizen-sensing devices seemingly invoke rights to data, to environments, to participation, and to breathable worlds. They promise to enable data citizenships that could redress the failure of rights. But the reinvention of rights requires worlds in which to take hold and become relevant. Data citizenships and citizen data do not solve the problem of partial rights and "citizenship contradictions."[83] Rather, they cultivate other strategies for making and remaking breathable worlds where data could become relevant. Through this process, citizen data can contribute to forms of action that reinvent rights, less as the pronouncement of universal, static, or fixed claims received in a uniform register, and more as attempts to build more breathable worlds within contingent, differential, and unequal environments. These practices express the right to monitor, the right to data, the right to participation, the right to environment, the right to experience, and the right to be political. But they do not assemble here as a straightforward implementation of a claim. Instead, they involve complex struggles to make worlds in which more just social and environmental conditions might be possible.

Such struggles can be generative of renewed conditions of citizenship in the making—as well as in the unmaking. As much as citizens and worlds are made and remade, they are also unmade and bound to unworkable conditions. Technologies of citizenship might need to be formed, transformed, and unformed.[84] These conditions are equally constitutive of data citizens, but are often overlooked by techno-optimistic narratives that would characterize these practices as effortlessly achieved. Making and remaking rights, citizens, and worlds is not an inherently liberatory process. Yet democratic engagement requires taking action that carries risks of uncertainty, disappointment, and failure.[85] Rather than transcend struggles to contribute to democratic life, citizen sensing and citizen data practices become interlocked with and co-constituted by these ongoing social movements.

How to Make Urban Worlds with Citizen Data

The right to data would seem to promise that more democratic and livable conditions could be realized through data collection. But as this discussion has suggested, it is not data for data's sake that would activate these changes. The right to data is not a linear sequence that activates the right to transform environments. By collecting data and identifying pollution hot spots, citizens support and mobilize projects to intervene within and reshape environments, along with the sociopolitical relations that contribute to conditions of (un)breathability. Rather than collecting data for regulatory compliance, citizen-monitoring practices struggle with the right to data as the right to make breathable worlds.

Figure 3.17. Dustbox installation in the Deptford Park area where Deptford Folk is active in proposing and developing changes in the urban realm. Photograph by Citizen Sense.

Deptford is an area with multiple construction projects underway within the context of heavy traffic, an incinerator, and river vessel pollution. At the same time, this area has a shortage of green space, a lack of sustainable transport, a high rate of poverty, and social and environmental injustice sedimented into neighborhood fabric. Within this context, a second monitoring area, Deptford Park, became a key site where citizen data contributed to ongoing efforts to transform transportation use in the area.[86] Several members of Deptford Folk, a community group working to improve parks in the area, installed Dustboxes to understand the effect of road transport on pollution levels and generally compare levels across the Borough of Lewisham.[87] Deptford Folk was established in 2015 to "improve parks and the routes to them." In a short span of time they began projects to improve local park infrastructure, plant trees, and undertake transportation pilots by temporarily closing streets to automobile traffic. People were interested in monitoring air quality to support and expand these ongoing initiatives. In addition, the South East London Combined Heat and Power (SELCHP) incinerator is located within the Deptford Park neighborhood, and residents were interested to see whether emissions from this site and nearby waste-transfer yards would show up in the citizen data set.

When MP Vicky Foxcroft tweeted about her call for a debate in the House of Commons about findings from the citizen data, Deptford Folk replied: "Let's debate but also let's take action: we're planting more trees in #Deptford as part of #Evelyn200. We're redesigning streets to reduce traffic and we're supporting people to take up cycling." This focus on "action" formed a key part of how Deptford Folk mobilized citizen data along with a multifaceted set of initiatives underway in the area. With ten times more cars on the road in London in 2012 than in 1949,[88] the need to address congestion was keenly felt. In advance of establishing where specific pollution patterns were occurring, the group was already in the process of testing transportation pilots with Lewisham Council to undertake traffic calming. They were hosting bicycle-repair workshops and preparing a larger funding application with a walking and cycling charity, Sustrans, to apply for Liveable Neighbourhoods funding from Transport for London (TFL) so that concrete improvements to transportation could be implemented in the area. They used their preliminary citizen-sensing data from the area to support their bid and to document how alternative transportation arrangements could benefit the area.

Yet citizen data did not unfold in an Enlightenment-style trajectory, such that once people had evidence of pollution they took action. Instead, Deptford Folk were already in the process of undertaking multiple initiatives to improve the local environment. These actions drew on data from planning portals, council documents, online data sets, websites, historic campaign activity (including John Evelyn's 1661 proposal to plant trees to improve air quality), among many other resources. Citizen-sensing technologies and data did not deliver a simple pathway to action. Instead, they became enfolded in these multiple and accumulating forms of evidence that variously supported attempts to improve transport and streets. In this sense, citizen data also joined up in a pluralistic way with multiple observations and other data sets from community group members, partner organizations, planners, city hall, and more.

In late 2017, Deptford Folk learned that their collaborative Liveable Neighbourhoods bid was successful, providing them with £2.9 million to undertake

Figure 3.18. SELCHP incinerator in background, with new bikeway in foreground; Westminster garbage trucks transport rubbish from West London to South East London to be incinerated at SELCHP. Photographs by Citizen Sense.

a larger study to develop sustainable transport in the area.[89] From 2017 to 2019 they collaborated with Lewisham Council and other organizations, as well as local residents, to review streets and transport in the area. As they worked to join up fragmented bicycle and pedestrian routes, connect green spaces, and create traffic-free areas, they frequently referred to the problem of air pollution as a key impetus for addressing excessive traffic in the area. Citizen data were mobilized in a supporting way to inform proposals for which streets to make traffic-free based on detected pollution hot spots. These proposals included addressing ingrained inequalities, including the uneven exposure to vehicle-based pollution in an area where many people do not own cars and are constrained in their ability to move around in an area without accessible walking, cycling, or public transport. With these proposals now having gone through community consultation, further work is still to be done to turn plans into interventions that could make this a more breathable urban milieu.

While these multiple community actions took place, Lewisham Council developed a series of air-quality actions, seemingly in response to findings from the citizen data but never explicitly acknowledged as such. The council expanded its regulatory monitoring sites to include a fourth station in Deptford as well as a fifth station that included a new monitoring supersite in Honor Oak Park. It developed a green-infrastructure fund for community groups, and it undertook no-idling campaigns and supported traffic-reduction initiatives.[90] Yet these efforts to address air quality were somewhat disengaged from the citizens and research groups who had worked to document, analyze, and propose different approaches to addressing air pollution. They were also relatively short-lived, since the council announced in late 2018 that it would cut its Strategic Air Quality Programme due to lack of funds and would instead focus on regulatory and statutory air-quality requirements.[91]

The breathability of worlds shows up here as the need to transform urban milieus based on felt and lived conditions. Citizen data were not the singular impetus for these transformations (despite the claims of technology companies). Instead, they supported but did not precede ongoing projects. Data were not the precursor to action, but they did reinforce the need for action. Data contributed to open-air instrumentalisms, along with the co-constitution of rights, citizens, and worlds in the making.[92] This more processual and pluralistic set of data operations demonstrates how citizen data become enfolded into rights claims not as fixed discursive expressions of individuals, but instead as prospective conditions constituted through material practices in the urban environment. In this way, rights and data are reinvented as instruments and toolkits that attempt to make and remake breathable worlds.

Figure 3.19. Deptford Park; tree selection guide used by members of Deptford Folk for a tree-planting campaign. Photographs by Citizen Sense.

By building more breathable worlds, people also hoped to connect air pollution to the health of urban environments and reduce the negative effects of pollution. In this way, the Ella Roberta Family Foundation,[93] named for a nine-year-old Black girl, Ella Kissi-Debrah, who died from asthma in this broader area of South East London, has also made a point of linking air pollution to the need to improve environmental conditions.[94] Ella and her family lived twenty-five meters from the South Circular, one of the busiest roads in London. Despite her multiple trips to ICUs due to asthma attacks, medical workers did not address how air pollution from roadways could be a factor in her asthma. The Ella Roberta Family Foundation was granted a second inquest to attempt to establish air pollution as the cause of Ella's death. The inquest was successful, and her death certificate now records air pollution as a cause of death.[95] Such an action could more directly establish the consequences of unbreathable worlds, where pollution literally constricts and collapses the lungs of those most vulnerable and most exposed to emissions sources. Here, the right to creature data includes the ability to categorically state that air pollution does kill people—nearly nine million worldwide every year, including 40,000 in the UK, and 10,000 in London alone.[96]

Residents have continued to protest this inattention to and neglect of air quality, notably with a campaign for cleaner air, "Let Lewisham Breathe," with Extinction Rebellion Lewisham.[97] In June 2019, protestors undertook a morning rush-hour disruption on major roads in South East London, including the South Circular (where Ella Kissi-Debrah had lived) and the A2 (where citizen data had documented air-pollution levels at six times WHO guidelines for twenty-four-hour averages). Protestors held signs that read "This Air Is Killing Us," "Lewisham Is Choking," "Deptford Needs Trees to Breathe," "Toxic Air Zone," and "R.I.P. Ella Kissi Debra 9 years old! Killed by pollution and asthma." These same calls to breathability gained renewed traction in 2020, with Black Lives Matter protests taking place across London and in cities worldwide and air pollution and deprivation exacerbating the impact of Covid-19, especially for ethnic minorities.[98] Such events assembled into a perfect storm of unbreathability, where as one protestor's placard paraphrasing Frantz Fanon noted, "We revolt because we can no longer breathe."[99]

HOW TO REINVENT RIGHTS

The need to make more breathable worlds is, more than ever, pressing upon citizens in the making. Breathable worlds materialize through collective engagements, political relations, and possibilities for constituting citizenships, rights, and milieus of exchange. The making and remaking of worlds with and through citizen data might work toward more breathable conditions. But this trajectory is not guaranteed. While writers such as Achille Mbembe call for a "universal right to breathe,"[100] the supposed condition of universality could work against the possibility of realizing—and reinventing—rights in highly differential conditions.[101] Instead, practices of combat breathing and situations that tune in to breath as an exchange and process that constitutes breathing entities, environments, and relations could generate sharper attention to the differential conditions that facilitate or impede the right to breathe. With these practices for making and remaking breathable worlds, it might also be possible to reinvent rights—to breath, data, participation, environments, and worlds—as a necessarily ongoing process of struggle.

From transportation experiments to the installation of a regulatory air-quality monitor in Deptford, the demolition of garden space, and the detection of pollution on the Thames, the effects and entanglements of citizen data developed in these collaborations between Citizen Sense and communities were complex and multiple. Sensors can format distinct modes of actionable data, yet they also mobilize forms of open-air instrumentalism that contribute to the making and remaking of urban worlds. As discussed in this chapter, citizens can generate data to support and create more just and livable cities. This is a particular way of understanding the right to make breathable worlds through the right of citizens to collect, analyze, and communicate data that dispute and question official accounts of problems such as air quality in relation to urban processes. The right to data then proliferates along with the right to clean air, the right to participate, the right to breathe, and the right to environment, as together they materialize as the right to make breathable worlds.

While data citizens form through multiple urban environmental data practices, they can also challenge and expand the usual ways of documenting and addressing environmental conditions. Indeed, one air-quality officer with whom I spoke about air-pollution levels in London stated that there was little that could be done about $PM_{2.5}$ levels in their borough, as the annual average of 19 µg/m³ varied by only +/-1 µg/m³ across their monitoring area. Particulate levels were seen to be attributable to pollution traveling from outside of the immediate area, or even from Europe or farther afield. From the expert's-eye view it might seem sensible to agree with the intractability of this problem, even though annual $PM_{2.5}$ levels of 19 µg/m³ are nearly twice the WHO's annual guideline of 10 µg/m³. Yet expert practices and infrastructures are here organizing the problem of air pollution in a particular way by assessing data sets according to annual averages as a measure of compliance (or not) with air-quality objectives. The numbers, which apparently record the facts of air pollution in London, will not budge, and so it seems we are stuck with the air we've got.

But data citizens can offer a diverging picture of urban air pollution by documenting differently granulated patterns and distinct city processes. Inevitably, when citizens work with "indicative" air-quality sensors that produce "just good enough data," multiple questions arise as to the accuracy of devices, the actors who can put forward evidence with sensor data, and the procedures and protocols that might be in place to ensure the validity of citizen data.[102] At the same time, an approach to air-quality monitoring that focuses on regulatory compliance offers just one way of investigating urban air pollution. Citizen air-quality monitoring can demonstrate a very different set of attachments and concerns, as well as ways of working with data and evidence that become practices of computing otherwise. Here, citizen data do not attempt to replicate or become an organ of expertise. However, they do constitute the problem of air pollution differently, which points to the plural worlds that converge and diverge through environmental crises such as air pollution. Data and data practices form distinct sites of collective inquiry, making, and remaking. These practices are also generative of distinct data citizens.

As this chapter outlines, certain ways of establishing the facts of environmental problems are treated as more credible than others, with significant consequences for how data citizens can make contributions, as well as how urban life is experienced. Ruha Benjamin suggests that empiricism often only works for some, since no amount of evidence will be accepted if the "facts" challenge the status quo or are presented by marginal or unauthorized voices. As she writes, "The facts, alone, will not save us."[103] Citizens who collect or analyze data might register new and significant observations, but these forms of evidence might not make a dent in political or regulatory processes. Those who are most affected by environmental pollution could be the least likely to be able to take up monitoring and have their data count. In this sense, rights to data are not easily configured through clear codes of access and use, since data might be "open" but only certain groups can mobilize or make claims with such data, often in relation to other data sources and with access to particular trajectories of power. As expressed through citizen data collection, the right to breathable worlds can be a project undertaken through struggle, but that falls flat if political environments and relations do not exist to build on that struggle.

The data citizen, in this sense, is not an automatically enlightened or empowered political subject. Indeed, it could be an ambiguous position, since data also require environments of relevance to take hold and have effect. Whatever accomplishments citizen data make in their observations, infrastructures, and collective experiencing, in order for them to realize less destructive environmental conditions these data also need to set in motion more just worlds that enable data to have an effect. Effectiveness, here, is less about the success or failure of data and more about the impasses that can arise when prevailing forms of political engagement break down or demonstrate hollow promises. The practices of data citizens can, in this way, work toward worlds where citizen data matter while also making more breathable worlds.

Citizen data practices attempt to support initiatives to make and remake worlds toward more breathable, just, and livable conditions. Data citizens and data citizenship materialize through these attempts to realize greater breathability by computing otherwise. But these practices are not just about the rights-based practices of preconstituted individual citizens. Instead, they involve searching out and making the conditions that would allow for collective exchanges across subjects and milieus. In order to realize the right to breathe, it might also be necessary to establish conditions to reinvent rights so that people can become citizens of worlds.

As noted in the introduction to this chapter, an increasing amount of legislation is being enacted to protect citizens' rights in relation to data, whether through tracking, the right to be forgotten, the right to open data, the right to transparency, and more. However, the generation of citizen data through citizen-sensing technologies raises the question of how data citizens and rights in the making are coextensive with worlds in the making. Rights to clean air might exist in some cities and countries, but these rights are frequently not observed. Citizen data practices have the potential to reinvent the terrain of rights—how they are formed, expressed, transformed, claimed, or abandoned. Such data practices form along with political subjects and collectives that are in search of more breathable urban worlds, but which rights do not fully support.

In this chapter I have examined data citizens and citizen data to consider on the one hand how data are produced in and connected to urban environments through sensors that monitor air quality, and on the other hand to study how citizens form environmental evidence that relates to their worlds of experience. While air-pollution monitoring instruments can be made to align, more or less, to detect a similar pollutant level in space and time, the actual uptake, use, deployment of sensors, and generation of data can veer into different directions when used by air-quality officials for regulation as opposed to residents observing and documenting changes in urban spaces. Not to attend to citizen data is to neglect urban dwellers' attachments to their cities, to the problems that matter in their lives, and to the practices whereby they document, analyze, and communicate evidence that speaks to their concerns. To make expertise the only register for producing legitimate data is to forgo and forget the importance of the environments that sustain data and allow data to have effect. It is also to suggest that an annual average calculated to comply with a regulatory guideline is the only way to organize the problem of air pollution—as well as the only way of considering how to create possible preventative and mitigating actions. To adhere to one official version of collecting data and forming facts is also to miss the question of which problems these facts pertain to and which worlds they sustain.[104]

It is possible *both* for experts' data indicating that annual-mean levels of $PM_{2.5}$ are 19 µg/m^3 *and* for citizens' data indicating specific patterns of elevated emissions when viewed as one-hour and twenty-four-hour data sets to be "accurate." Each of these forms of data takes hold and gains relevance within distinct worlds that can offer diverse responses to environmental problems. If a more pluralistic ontology of data and data practices were to be realized, then both—and more—of these creatures of data would need to be recognized as relevant to inundated urban habitats. Indeed, the very qualities of expertise could begin to shift and respond along with the environmental conditions that are meant to be governed toward more collective projects. Here is where data citizens materialize as figures constituted not just through digital technologies or observational practices but also through their concern for relations and communities on behalf of which evidence would be mobilized.

No singular figure of the data citizen concretizes here. These are, as Berlant has suggested, proliferating forms of citizenship, since they are tied to the worlds that are endured, narrated, created, and hoped for. Proliferating modes of citizenship are indications of different experiences that will inform how rights in the making are taken up, if at all, and the struggles they produce. Here, the right to data and the right to make breathable worlds are co-constitutive. The right to clean air is not simply about meeting a regulatory threshold for criteria pollutants; it is also about transforming the urban processes and milieus that are grinding away at conditions of breathability. These affective engagements are productive of different ways of being in, as well as making and remaking, worlds.

TOOLKIT 4

PHYTO-SENSOR TOOLKIT

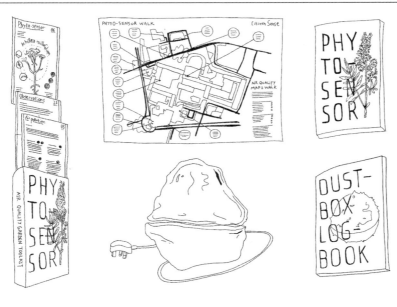

Phyto-Sensor Toolkit developed by Citizen Sense for developing air-quality gardens and monitoring air quality. Illustration by Sarah Garcin; photograph by Citizen Sense; courtesy of Citizen Sense. This toolkit can be found in a more extensive form online at https://manifold.umn.edu/projects/citizens-of-worlds/resource-collection/citizens-of-worlds-toolkits/resource/phytosensor-logbook.

Chapter 4

MULTIPLE CITIZENS

How to Cultivate Relations

In the Square Mile—the financial center of London, where narrow streets, Victorian buildings, and Roman walls collide with glass-and-steel skyscrapers—unusual gardens are springing up. Planted in corrugated steel pipes and granite troughs, vegetation grows at roadside edges and adjacent to construction sites, where it absorbs, exchanges, and filters air pollution. Community groups and residents' associations have developed these air-quality gardens to draw attention to elevated levels of air pollution in central London. The gardens demonstrate how plants sense, absorb, and capture gases and particles from the air while making and remaking urban atmospheres.

As it turns out, environmental sensing occurs through more than just technologies in their usual configuration. Environmental sensors are well established as analog or digital instruments for measuring pollution and other disturbances. But more-than-human organisms also sense and communicate environmental processes. Bioindicators and sentinel organisms, from birds to marine life, detect and respond to changes in their milieus.[1] Such exchanges across organisms and environments demonstrate how sensing involves multiple entities and worlds of experience. More-than-human modes of sensing can also become more or less evident through digital sensors that work with and alongside organismal detection. These practices present yet another way of computing otherwise by reworking relations with other organisms.

In this way, plants perform as sensors and measurement devices that detect environmental changes. Some plants signal the presence of particular pollutants in environments through their leaves and growth patterns. Other plants are especially effective at taking up pollutants by absorbing gaseous substances through their stomata, drawing in heavy metals from the soil through their roots, or channeling and depositing particulates on their leaves. Plants, lichens, and mosses

Figure 4.1. Installation of Dustboxes in air-quality gardens at the Museum of London. Photographs by Citizen Sense.

sense and register air pollution through changes in their structure, distribution, and prevalence. As bioindicators, plants do not typically provide an instantaneous measure of pollution levels but rather signal relative gradients and relationships that become evident within environments over time. In sensing pollution, plants materialize other registers of experience that involve more than one organism responding to stimuli. In other words, plants incorporate, signal, and respond to environmental pollution through relational modes of sensing.[2]

Although vegetal participants are involved in sensing, absorbing, capturing, recycling, and reworking pollutants, within the wider context of "citizen sensing" more-than-human organisms do not typically register as "citizen sensors." In an attempt to expand the usual human-to-machine configuration of citizen sensing,[3] this chapter takes up more-than-human ways of sensing environments to consider how multiple other entities contribute to projects to make more breathable worlds. By focusing on the construction of two air-quality gardens as sites of vegetal and digital pollution sensing, I investigate how citizens and sensing transform through these more-than-human relations and world-making practices.

In constructing these gardens, Citizen Sense worked with several collaborators, including the Museum of London, a landscape architect, and the air-quality team at the City of London Corporation, to develop and test air-pollution gardens that were co-located with air-pollution monitors. We also engaged with museumgoers in a workshop and walk to develop and refine a Phyto-Sensor Toolkit that provided how-to instructions for developing air-quality gardens. In describing the planning, design, construction, and tool-kitting of air-pollution gardens developed in the financial center of London, this chapter documents how plants became evident as sensing organisms that capture and mitigate air pollution, and how these other modes of relational sensing recast environmental experience.

Air-quality gardens sense and transform polluted urban air through vegetation that traps and absorbs pollutants. Such gardens cultivate the bioindicating and mitigating qualities of plants to address the effects of air pollution. In this context, citizen-sensing practices materialize as a composition of air-pollution plants and digital sensors, along with residents, workers, cultural institutions, and local government agencies. The "citizen" operates here less as a universal or individual human agent and more as a project of activating sociopolitical relations across pluralistic registers. Adding to the proliferating list of citizens discussed throughout this study, I suggest these are *multiple citizens*, formed with and through other entities and worlds in the making. Citizens, in other words, form and exist in multiples, as entities involved with many other entities.[4] By investigating these more-than-human sensing relations and collectivities, I consider how exchanges across subjects and worlds make and remake possibilities

Figure 4.2. Dustboxes and Phyto-Sensor Toolkits as part of an air-quality garden workshop at the Museum of London. Photographs by Citizen Sense.

for being and becoming citizens. To engage in this area of inquiry, I consider how more-than-human organisms take up and express changes in their environments. I look especially at how more-than-human air-quality sensing contributes to distinct practices for making and remaking subjects and worlds toward more breathable conditions. And I examine how these affiliations multiply and recast the practices and worlds that could be constitutive of citizenship.

This chapter investigates how practices of sensing environmental pollution generate different articulations of citizens and experience when extended through more-than-human affiliations. Digital sensors might tune humans in to more-than-human sensing. But more-than-human sensing does not simply replicate what a digital device indicates. Instead, these other modes of sensing signal how relations and exchanges within milieus sediment over time. Processes of sensing pollution through organisms rework the boundaries of subjects and environments. What surfaces through these sensing practices is not just measurable levels of pollution but also the ongoing and accumulative effects of pollution in lived environments.

By engaging with organismal contributions to environmental problems, this chapter explores how different approaches to sensing generate multiple citizens and citizenships. Citizens and citizenship in this sense are not sole attributes of human actors. Instead, they are conditions activated through exchanges with other entities and milieus. The "multiple" is as much about multiplying entities as multiplying relations that co-constitute worlds and ways of being and becoming citizens. When multiplied across entities and milieus, citizenship becomes a capacity of differently activated worlds. Such a shift in focus reworks the subjects, practices, and worlds for experiencing and addressing environmental pollution.

Throughout this study, I have discussed how the conditions of being and becoming citizens depend upon the breathability of worlds. As a form of breathability and breathing, sociality can constitute what Ashon Crawley calls an "openness to worlds."[5] Such openness is part of "a critique of the ongoing attempt to interdict the capacity to breathe," which cultivates collective and aesthetic exchanges that come "to constitute community."[6] Crawley works against the racialized and reduced modes of being that form through closed and airless approaches to subjects. Breathability involves exchanges across subjects and worlds, and these exchanges can form and transform subjects and worlds beyond limited modes of being.

Such openness to worlds also requires sociality and exchanges with and across multiple more-than-human entities and milieus. Robin Wall Kimmerer conveys how breathing involves and generates conditions of reciprocity across multiple entities when she writes, "The breath of plants gives life to animals and the breath of animals gives life to plants. My breath is your breath, your breath is mine."[7] These exchanges are more than physical interactions. They are also constitutive of political subjects and modes of governance, forming what Kimmerer refers to as a "democracy of species."[8] In this way, pollution surfaces not simply as an event to observe and detect because it interrupts or obstructs breath. Instead, it is a problem that activates expanded subjects, relations, practices, and collectives for working and reworking environments.

Citizens form through these processes, exchanges, and struggles, where breath-ability involves a collective searching toward open air that constitutes political life. This chapter investigates how to tune in to collective respiration with more-than-human entities as a condition of making citizens and worlds. As in earlier chapters, I explore how making breathable worlds gives rise to distinct citizens and citizenships. Still, here I emphasize the more-than-human exchanges that inform political life. Air-quality gardens attempt to improve breathability by de-creasing air pollution and remaking urban conditions by engaging with vegetal sensors. They potentially contribute to different ways of constituting citizens and citizenship. The incorporation of more-than-humans into projects of politi-cal engagement could expand the breathability of worlds by pluralizing the exchanges with and through which community is constituted. These citizenly reconfigurations could contribute to the right to be in relation with environ-ments and multiple other entities, carrying on from the last chapter. Yet such projects also demonstrate how difficult it is for some citizens to address air pol-lution, where emissions often continue unchecked, and remedial interventions that require the contributions of multiple organisms become a stopgap measure for making more breathable worlds.

Citizens form in multiples: as a proliferating range of political engagements, as a differential cast of human and more-than-human subjects, and as a plural-ity of relations and worlds in the making that unfold in conditions of strug-gle. Focusing on projects to make air-quality gardens in London, I consider how citizen-sensing practices exceed a more typical or scripted enactment of citizen-ship through devices. Instead, these practices spill over into urban projects that enlist monitoring but work toward other configurations of citizens and worlds. I turn now to examine how citizens multiply by incorporating more-than-humans into projects of sensing air pollution. I then document how the Citizen Sense research group engaged with Londoners and London institutions to build air-quality gardens and test different approaches to sensing with vegetation through a Phyto-Sensor Toolkit. In this discussion, I consider how citizenship materializes across multiple entities through how-to practices that include how to construct air-quality gardens, how to transform citizen sensing through citizen gardening, how to aerate a Phyto-Sensor Toolkit, and how to multiply citizens in worlds. More-than-human modes of sensing surface here as propositions for how to expand environmental subjects and relations toward more breathable worlds. When constructing an air-quality garden, you might find that relationships across citizens, environments, and multiple organisms can be reworked in more reciprocal and expansive ways by computing otherwise.

SENSING MULTIPLE CITIZENS

Vegetal sensing, which the Phyto-Sensor Toolkit activates, is a method that offers different ways to tune in to environmental processes. Unlike digital sensors, plants do not typically produce real-time measurements of environmental variables, whether temperature, particulate matter, or ultraviolet rays.[9] Vegetal sensing also does not easily translate into the standard taxonomies of human sensing. Instead, such sensing unfolds through distributed and mutual responses, incorporations, and transformations in environments. As one mode of more-than-human sensing, bioindication is a process whereby environmental pollution registers in the bodies, distribution, growth, frequency, and relations of organisms. Organisms express physiological or other observable changes that can also indicate the accumulation or duration of pollution events—or even their possible recovery from pollution. Changes in organisms can signal changes in environments. These more-than-human and multiple modes of organismal sensing show how pollution is a relational event that opens toward ecological configurations of entities, which also become sites of political potential.

Bioindication is one such mode of sensing that demonstrates how environmental subjects materialize as they experience the lived effects of pollution.[10] Here, practices of monitoring and measuring pollution do not rely on instruments, in the usual designation, to document quantitative values of atmospheric pollution. Instead, organisms become gauges by indicating the spread and accumulation of pollution in particular sites, as well as approximate levels of pollutants, through their form and growth patterns.[11] Often referred to as a "qualitative" mode of monitoring environments, bioindication expresses how environmental processes such as pollution transform particular organisms in a nonlinear and yet accumulative way.

Bioindicators are species that express certain qualities of ecosystems, especially in relation to pollution. Biomonitoring can be undertaken by studying organisms in their environments or through transplant studies, where organisms are brought into environments to understand how they react to pollution.[12] Indicators are often compared to an index, which establishes a set of protocols for reading and describing the environmental condition that the indicator is expressing. The qualitative aspects of bioindication are a unique aspect of this mode of monitoring. Yet they can align with quantitative monitoring if undertaken systematically through the use of indices.[13]

Bioindication highlights how pollution influences the material transformations of organisms in their habitats. It can map onto numerical indices of pollution, and it can also provide a way to understand the possible "health," "vitality," and even "luxuriance" of organisms by providing a more dynamic, ecological, and complex understanding of pollution.[14] Here, pollution is less about a numerical value and more about an ongoing set of transformative effects that can rematerialize and remake environments. Bioindication requires homing in on the effects of exposure over time, not necessarily as an instant measure of air quality but rather as durational materializations of pollution that affect wider ecologies. Different ways of "taking measure" and attending to the expressions of pollution through organisms can reorient attention from isolated variables to experiences and relations. Such a shift in focus can remake the designation of environmental problems (along with collectives and actions) by differently attending to the relations that pollution affects.

World Making and Phytosociological Associations

These vegetal operations, then, establish how to sense is to transform. By making and remaking atmospheres and soil, plants not only process pollution but also contribute to projects of multispecies world making.[15] World-making projects have been discussed not just through pragmatist philosophy but also through the work of feminist technoscience and Indigenous scholars.[16] Writing about world making in her discussion of the Matsutake Worlds Research Group, Anna Tsing notes that "each living thing remakes the world" through "multiple time-making projects" that transform landscapes.[17] These worlds are in the making through the activities of multiple entities as they contribute to forming distinct environmental conditions.[18] Such conditions indicate how organisms are in correspondence with and diverging from numerous other entities, responding to environmental events in different ways and across variable durations.

Vegetal forms of more-than-human sensing can show how these organisms experience pollution and form as distinct environmental subjects and worlds through responses to environmental pollution. Environmental pollution often registers through measurements that exceed acceptable levels or health effects and challenges made to regulators and polluters. Yet pollution also becomes evident through ongoing transformations of organisms and environments. Such modes

Plate 1. Citizen monitoring of gas emissions near fracking infrastructure; pro-fracking signs at vacated home site known to have polluted well water in Dimock, Pennsylvania. Photographs by Citizen Sense.

Plate 2. A Speck particulate-matter monitor with DIY weatherproof housing; pro-fracking sign protesting zoning regulations of fracking industry, Pennsylvania. Photograph above by Citizen Sense; photograph below by anonymous participant; courtesy of Citizen Sense.

Plate 3. Signs alert passersby to air pollution on busy road in London; testing citizen-monitoring technology during a walk in New Cross Gate, South East London. Photographs by Citizen Sense.

Plate 4. Installation of Dustbox in the Beech Street regulatory air-quality monitor. This monitor, sited at the edge of a traffic tunnel and near the Barbican Tube Station, regularly registers high levels of nitrogen oxides and particulate matter. Photograph by Citizen Sense.

of environmental sensing highlight how entities and environmental affiliations are affected by and transform through pollution. While in some ways bioindication might become legible through correspondence to an index that provides a systematic way of observing pollution, in other ways it gives rise to more open-ended engagements of incorporating and responding to environmental change. These open-ended or open-air instrumentalisms signal the distributed capacity of organisms and environments to generate different propositions for being together.

As I have discussed throughout *Citizens of Worlds*, environmental sensing—and the practices and entities involved in monitoring—can make environmental struggles matter in distinct ways. Measuring air quality with regulatory monitors is one way to sense environments. Attending to pollution as it accumulates in organisms, bodies, and ecologies is another way to tune in to environments. These different modes of sensing generate distinct forms of evidence and relevance. They also show how various entities and collectives work through and are individuated by pollution. As Gilbert Simondon has developed the concept, individuation refers to how entities are in-formed in relation to each other and their milieus.[19] Subjects, relations, and milieus all have the potential to transform and are not pregiven, although they can be in-formed by sedimentations and inheritances. Here, a "human" is not a fixed entity and can shift in relation to different relations and milieus.[20] So too might a "citizen" be encountered as an entity that changes through the collectives and worlds that affect it and to which it contributes.

In this way, subjects—including political subjects—are bound up with multiple other organisms and environments. Research focusing on the symbiotic characteristics of organisms suggests that even the notion of an organism as an individual is fraught with problems. Scott Gilbert, Jan Sapp, and Alfred Tauber have suggested through their work on symbiosis not only that "we have never been individuals" but also that "we are all lichens."[21] Indicator species such as lichens, plants, and many other organisms can provide a way to rethink environmental change through relational distributions of subjects. Such an approach can generate a less anthropocentric rendering of a changing planet, while connecting environmental subjects to earthly conditions that influence them.[22] Indeed, if lichens such as *Umbilicaria* are present, Kimmerer notes, "you know you're breathing the purest air. Atmospheric contaminants like sulfur dioxide and ozone will kill it outright. Pay attention when it departs."[23] The disappearance of bio-indicating organisms can signal the contamination of the atmosphere that affects multiple other organisms and relations, which in turn could also be negatively affected by pollution.

Such a remaking of subjects through environmental affiliations has consequences for how to constitute political subjects and engagements. This distributed and multiple rendering of a subject is far from the universal and static figure that propagates through conventional humanism. A subject necessarily includes all that sustains it. Bioindication is not just an organism signaling the presence of pollution. It also expresses broader environmental engagements that influence earthly inhabitations. Organisms can draw attention to conditions of what Tsing calls the "more-than-human sociality" that informs environmental processes.[24] For instance, a forest encompasses the bioindicating characteristics of individual organisms, and materializes and sustains the more-than-human social worlds that are made through these organisms.

More-than-human sociality aligns with the "community approach" to bioindicator-based monitoring, "where indicator values are not assigned to single species or communities, but to groups of species with similar ecological requirements."[25] Rather than monitoring individual entities or variables, these more collective approaches can document how ecological relationships are expressive of environmental pollution and change. The shared "phytosociological associations" and "multidimensional relationships"[26] of indicator organisms are an area that has hitherto been somewhat under-studied, however. While scientific texts might chalk this omission up to the lack of a systematic methodology for evaluating community indicators, fledgling methods have been developed that, for instance, indicate air pollution through phytosociological relationships such as epiphytic organisms that grow on trees. Acidity and toxicity, the availability of light and water, and the abundance of nitrogen affect these organismal relationships. Similar to the *Umbilicaria* lichen, the presence or absence of epiphytes signals broader processes at work in the composition and decomposition of environments.

In other words, bioindication as an expression of ecological communities can be a way to get a sense of the health of environments, since "the replacement of a community by another one can be considered a clear indicator of environmental change."[27] Christelle Gramaglia and Delaine Sampaio da Silva suggest that "sentinel organisms" can signal conditions of pollution while at the same time demonstrating how an ecosystem such as a river operates as a "collective entity."[28] These more-than-human modes of sensing involve other communicative exchanges, as well as distinct expressions of sentience, intelligence, and "meaning" that require connections to numerous ecological operations.[29] Rather than designating how a stable nature might be sensed, or bioindicated, by other organisms, more-than-human sensing is instead expressive of multiple environmental inhabitations and relations.[30]

Vegetal sensing draws attention to how environmental subjects take account of and form distinct experiences of their worlds.[31] Connections across organisms and environments are continually remade through the accumulation and dissipation of pollutants. More-than-human sensing signals relationships across organisms, the composition of ecological communities, and the effects of their inhabitations. Such modes of sensing serve as provocations for how to characterize, engage with, and act on environmental problems. If "practice imposes upon its participants certain risks and challenges that create the value of their activity,"[32] as Stengers suggests, then more-than-human modes of sensing indicate how risks and challenges cascade across pollution, bodies, environments, and relations. Practices of vegetal sensing recast how pollution becomes evident as well as how it might be acted upon. By incorporating multiple entities into practices of citizen sensing, it could be possible to rework conditions of political possibility while generating alternative strategies for making more breathable atmospheres.

The Political Work of Multiple Citizens

Air-quality gardens draw attention to how more-than-human relations contribute to breathable worlds. Vegetal sensing offers other practices for inhabiting environments and for constituting citizenship. These are multiply constituted modes of citizenship that, drawing on Tsing, attempt to "make common cause with other living beings" by engaging with more-than-humans as "political work."[33] Such political work involves reciprocity and exchange—of sensing, breathing, making, and remaking. Possibilities for more expansive environmental politics surface through tuning in to the world-making projects of other organisms. Air-quality gardens can be spaces for cultivating and pluralizing sensing subjects. They work and rework environmental pollution toward other earthly inhabitations.

Multiple citizens take shape through the "collective potential" generated by these different affiliations, exchanges, and ways of parsing environments.[34] Sensing processes and relations give rise to multiple citizens and citizenships that remake environmental struggle and politics. Here, an environmental citizen is not the familiar figure of a responsible consumer–subject amenable to behavior change but rather an opening into other practices and relations of more-than-human sociality and political engagement.

However, the multiplications of citizens and citizenship do not signal a mere concatenation of actors. Instead, they are sites of struggle. Struggle here is not merely or necessarily a political contest that takes the form of "debate." Instead, it forms through environments at saturation points, species loss, disconnection from land, and contaminated ecologies. The struggles that shape the political

Figure 4.3. Air-quality garden installations at the Barbican in the City of London. Photographs by Citizen Sense.

capacities and designations of subjects are multiple. Yet this plurality might also move toward "more different ways of being in relation," as Berlant suggests in the first epigraph to this book's Introduction. In this way, citizenship is always in process and "produced out of a political, rhetorical, and economic struggle over who will count as 'the people.'"[35] These struggles extend not just to who counts as the people as a collection of humans but also to the possibilities for being in relation with other entities and environments that might make for less destructive worlds. The demos that materializes here necessarily extends to the incorporation of more-than-humans and environmental relations as part of what makes these worlds breathable. Such pluralizations become sites of possibility by transforming the conditions and relations that could be activated through citizens of worlds.

In this regard, Kim TallBear points to the need to "indigenize" fields such as science studies and animal studies, which produce human subjects detached from environments and organisms and separate the organisms they study from their milieus. Drawing on American Indian metaphysics, TallBear seeks to "extend the range of nonhuman beings with which we can be in relation," which in turn would inform the ecologies we inhabit and the environmental subjects we become.[36] Here, TallBear points to how practices for indigenizing environmental subjects can transform social and political engagements by attending to earthly relations. Subjects are not pre-constituted as free-floating cognizers of worlds. Instead, worlds inform and co-constitute subjects. Or as Jennifer Wenzel writes regarding Frantz Fanon's *Wretched of the Earth*, "decolonization demands . . . a shift in the valuation and disposition of nature,"[37] not as an external entity against which to define humans but as a reworking of humans and more-than-humans toward less extractive and more reciprocal relations.

In this inquiry into how citizens of worlds form and are sustained, multiple modes of citizenship and "membership" materialize. Writing about historical restrictions to Indigenous citizenship, Leanne Betasamosake Simpson explains how nationhood is remade and differently lived within Indigenous worlds, less as a territory of violence and hierarchy and more as a web of relations. Indigenous nations are in relation with plant nations, animal nations, and many other more-than-human nations. A nation is an "ecology of relationships" that distinctly constitutes power and governance.[38] Citizens and citizenship are then reconstituted through connections to multiple other entities and milieus, which are formative of political subjects. Human citizens form in relation to councils of plants, animals, and atmospheres, among countless other collectives. They make contributions along with vegetal citizens and plant nations, which form other possibilities for collective affiliation and environmental inhabitation.[39]

Such formations of environmental citizenship are an attempt to rework political subjects through multiple relations. Plants, among many other organisms,
open into other inhabitations and durations.[40] They draw attention to environmental attachments and formations of subjects through the signaling, incorporation, and transformation of pollution. The next section describes how Citizen
Sense engaged with collaborators to construct and test air-quality gardens. On
one level, this initiative could be seen as an idiosyncratic gardening project. On
another, through its attention to how multiple organisms respond to air pollution,
it offers an invitation to cultivate practices for working with more-than-human
citizen sensors toward more breathable worlds.

CULTIVATING MORE-THAN-HUMAN WORLDS

Sensors would initially seem to activate and format citizens as collectors of data.
Yet such a diagram of action is never so straightforward. As described in the previous chapter, communities in South East London often mobilized data to support and extend sustained efforts toward environmental and social justice. But
practices for tuning in to urban environments were not linear expositions of
evidence. Instead, they were formed through ongoing experiences that often drew
on data in more provisional ways. Citizens proposed multiple actions that they
formulated in advance of, contemporaneous with, and based on evidence from
air-quality monitoring. These proposals included responding to findings about
the important role of vegetation and green spaces in mitigating air pollution by
improving green infrastructure, planting trees, and protecting green space. Indeed,
some of the lowest levels of air pollution in this citizen-monitoring network were
found in an enclosed garden on a pedestrian street. While citizens' proposals did
not take the direct form of air-quality gardens, they did advocate for tree planting
and green-space preservation. These proposals demonstrated how citizen sensing
becomes involved with more-than-humans when attempting to cultivate more
breathable worlds.

While we were in the process of finalizing our collaborative monitoring in
South East London, in June 2017 the Museum of London wrote to Citizen Sense
to ask if we would like to contribute to a project to develop air-quality gardens
and test our sensors at the museum.[41] Working along with the City of London
Corporation and supported by funding from the Mayor of London Low Emission
Neighbourhood scheme, we helped develop the gardens as part of a broader
initiative to investigate and implement projects for improving air quality. Related
London-wide initiatives included transportation experiments (as discussed in the

previous chapter with Deptford Folk's successfully funded pilot project), no-idling campaigns, and green-infrastructure installations.

There has been considerable debate about green infrastructure within air-quality research communities. Some air-quality researchers see a focus on vegetation as distracting from the key problem of removing emissions at the source. In contrast, others suggest that a wide range of approaches should be explored to improve air quality. Here, the question of practicability also arises: citizens might be less well placed to redesign transportation infrastructure, for instance, than to plant air-quality gardens. The difficulty of shifting to less fossil-fuel-intensive transportation proves intractable across all levels of governance, both in the UK and farther afield. Citizens take up projects such as air-quality gardens because other efforts to address air pollution have failed or because they feel this is an intervention that is within reach. Practices for addressing air pollution become informed by these more or less feasible struggles to transform cities toward more breathable worlds. Thus the relative feasibility of political engagement directly impacts organisms not just through biodiversity collapse and extinction but also through the distinct types of work in which more-than-humans become enrolled as they process and transform the residues of fossil fuels and the pollutants from extractive economies.[42]

With the air-quality garden installation, the Museum of London was interested in developing a demonstrator project to facilitate collective research into air quality. We considered how the garden could include plant installations, air-quality monitoring, and public engagement and events. Citizen Sense contributed to these aspects of the project by providing suggestions for air-quality plants that would bioindicate and mitigate air pollution, installing Dustbox sensors in gardens and other nearby locations, hosting a workshop and walk with museum-goers to investigate air-quality vegetation and gardens, and developing an openly available Phyto-Sensor Toolkit for the wider development of air-quality gardens. In addition to our contribution to the gardens, Grow Elephant, a London-based landscape architecture firm that had developed temporary community gardens and garden clubs within South East London, built two garden planter structures and installed the air-quality plants. After a relatively rapid development process in the summer of 2017, the air-quality demonstrator gardens opened at the museum in September as part of its *City Now City Future* exhibition, with one garden at the entrance to the museum and one situated at a walkway intersection.

I describe in more detail the process of how we constructed these air-quality gardens below. Yet it is important to note that this project did not materialize as a singular endeavor. Instead, the Museum of London's air-quality garden exploration

Figure 4.4. City in Bloom and air-quality garden initiatives developed through the City of London with local residents. Photographs by Citizen Sense.

took place within a broader context of multiple other community air-quality initiatives underway in the City of London. These initiatives tended to draw on participants from demographics that were very different from South East London or rural Pennsylvania. Residents in the City of London are primarily (although not exclusively) from more privileged economic backgrounds and work in (or have retired from) higher-paying professions. But similar to South East London and rural Pennsylvania, there was considerable variation in the types of participants engaged in environmental monitoring.

Residents at the Barbican and Golden Lane Estates had also developed several air-quality gardens in the area. Through a "City in Bloom" initiative, residents compiled a working pamphlet on "Why Plants Can Help Improve Air Quality" to share scientific research on the topic. Based on their research and garden projects, they also compiled *The Clean Air Gardens*. This self-published book includes suggestions for air-quality plants and documents nineteen gardens that they installed throughout the Square Mile.[43]

One of these temporary or "pop-up" gardens at Moor Lane involved a collaboration with landscape architecture students who had recently set up their own design firm, xmpl. The residents and designers obtained donations from construction companies developing transport infrastructure and office complexes to create a unique intervention to address air pollution.[44] Because construction companies at times want to be seen as "good citizens" while undertaking highly disruptive projects, residents and community groups in this area of intensive urban development successfully obtained donations and small amounts of funding to "green" the area. Greening initiatives can often exist in this blurry zone of political engagement. They attempt to make more livable environments, while working within prevailing conditions of urban development to make situated interventions.

Community groups in the area also organized and ran bird walks, bat walks, and tree walks to survey species and introduce people to different ways of looking at and listening for multiple organisms in the Square Mile. They were involved in citizen-science initiatives through the City Nature Challenge, where they surveyed and recorded organisms in the area using iNaturalist to determine relative levels of biodiversity. One local resident led an extensive survey of vegetation in the area to develop a herbarium of plant pressings for the Natural History Museum archive. She located her collection of over two hundred plants on a Google map to document the urban habitat in which the species grew. London plants, as one Barbican resident noted, are never just randomly placed but instead always carry stories about how they came to occupy particular areas. Fireweed is a plant that is well known for spreading in former bomb sites, where it occupies the ruins of once-cratered urban landscapes. Urban ragwort had seemingly been transplanted from Italian volcanoes to the Oxford botanical garden,

Figure 4.5. Construction on the perimeter of the Barbican; internal gardens at the Barbican in the City of London. Photographs by Citizen Sense.

where it traveled up train lines into UK cities. This citizens' herbarium formed a record of flora in an area where green spaces were often transformed or crowded out under constant pressure from development.

Along with these monitoring practices, a group of Barbican residents had set up a diffusion-tube air-quality study to document nitrogen dioxide levels in this central area of London. The Barbican Estate experiences poor air quality because of its location within a dense urban area that suffered from traffic congestion, hard surfaces, and few green spaces or parks. In 2015 they worked with Mapping for Change, a research agency based at University College London, to analyze monitoring data and document the study in a report.[45] They found that pollution levels were especially high on the edges of the estate that adjoined busy roads. However, they also found that nitrogen dioxide levels could be elevated even in garden courtyards when high winds circulated from particular directions.

These monitoring efforts informed a wide range of activities to address, mitigate, and raise awareness about the problem of air pollution, some of which were undertaken in collaboration with the City of London.[46] For instance, the diffusion-tube monitoring study further reinforced the clean-air gardens initiative as a concrete strategy for addressing air pollution. It was in this context of local air-quality monitoring projects, biodiversity surveys, and installing clean-air gardens that Citizen Sense undertook a practice-based investigation into how two air-quality demonstrator gardens could activate expanded relations with more-than-human sensing and work toward more breathable worlds.

How to Construct Air-Quality Gardens

The process of setting up two air-quality gardens at the Museum of London involved much more than locating plants within planters. The demonstrator gardens drew on local initiatives, from monitoring to gardening, while communicating with and connecting to more dispersed publics about projects for addressing air pollution. The demonstrator gardens did not operate as community gardens in the traditional sense of a sustained cultivation space. Nor were they a project for providing food or contributing to (or reforming) disadvantaged communities through gardening, a topic that researchers have discussed as having complex social and political implications.[47] While residents were involved in looking after the nineteen clean-air gardens they installed, Grow Elephant supervised the demonstrator plots during their temporary installation. In this way, the gardens provided a test site for people to learn about air-quality plants, assess how to develop their own gardens, and consider strategies for mitigating air pollution. The gardens also provided a space to study the possible effects of air pollution on plants, while comparing pollution levels from the Dustbox particulate-matter sensor.

While the Phyto-Sensor Toolkit began with the demonstration gardens, it also encompassed a workshop and walk, a physical and online toolkit, and a broader investigation of practices for working with more-than-human sensors.[48] *Phyto-sensor* is a term that the Citizen Sense research group uses to describe the vegetal entities and processes that sense and respond to environmental pollution. The gardens, events, and toolkit explored how plants continually sense and change environments. Because the sensing, bioindicating, and mitigating characteristics of plants are not always immediately evident, each demonstrator garden had a placard noting that the plants had been selected for their "special properties, either absorbing or channeling pollutants, or indicating particular pollutants with changes in their physiology and appearance."[49]

The material registers of pollution became more or less evident through the garden, toolkit, events, and different engagements with vegetal sensors. Yet as demonstration spaces, the gardens were zones of inquiry that were focused less on identifying immediate changes and more on learning about more-than-human sensing and the relations it activates. Plants in their milieus became sensors and teachers.[50] To this end, the toolkit documents how some plants are especially effective at taking up pollutants. These other organismal and vegetal ways of sensing environments served as a basis for investigating how to cultivate more breathable atmospheres, especially by developing planting scenarios that could be installed in other areas.

We selected plants for the gardens that included species that could bioindicate the presence of pollution, absorb or trap pollutants, or accumulate and transform toxins. Many plants work across these different ways of sensing and responding to pollution. The garden included bioindicating plants such as snowberry (*Symphoricarpos albus*), which is sensitive to ozone and will show leaf injury and impaired growth when ozone is present.[51] Snowberry can also take up heavy metals such as zinc and iron from the soil, but it will show signs of damage and impaired growth if it accumulates these contaminants. Thus the garden included plants that demonstrate how contaminants persist and transform organisms and their environments, often with deleterious effects.

Many of the plants in the air-quality garden trap or channel particulate matter through deposition and dispersal. Plants with hirsute or broad leaves can act as surfaces to capture airborne particles. When planted as a hedge or green screen, plants such as yew (*Taxus baccata*) can catch particles that wash into the soil during rainstorms. At the same time, many plants are respiring and contributing their own mix of materials to environments, including oxygen and other gases. Yew emits low levels of biogenic volatile organic compounds (BVOCs), which can combine with nitrogen oxide to form ozone. Many urban plants and trees emit

Figure 4.6. Air-quality garden installed at the entrance to the Museum of London; Phyto-Sensor Toolkit used as part of a walking tour of air-quality gardens to identify plants and learn more about how they respond to air pollution. Photographs by Citizen Sense.

BVOCs. For this reason, simply planting more vegetation because it is "green" does not necessarily improve air quality, and can instead exacerbate pollution in areas with high levels of nitrogen oxide that can react with BVOCs.

Plants bioremediate air pollution in ways that are not simple actions of scrubbing and removing the residue of fossil fuels. Instead, they absorb some pollutants while emitting others. They capture particles from the air while carrying them into the soil. They phytoremediate contaminants in the soil but often can only do this when in relation with other plants. They bioindicate the presence of pollutants such as ozone, but this form of sensing and signaling can also demonstrate that plants are likely to die. The air-quality gardens and Phyto-Sensor Toolkit document these bioindicating and mitigating characteristics, showing how plants such as yarrow (*Achillea millefolium*) can colonize soil in urban areas and provide a surface for particle deposition. Yet yarrow can also become damaged by ozone, a condition that becomes more or less severe in relation to the other species with which it grows.

Because these modes of sensing register through the form, growth, and distribution of plants and are typically more evident when they grow over time in environments, we also set up Dustboxes to monitor real-time pollution levels at several locations within and near the gardens. This Dustbox installation was also a process of consolidating our sensors for wider circulation as a sensor library, which might be used as a mobile and DIY air-quality monitoring infrastructure for installation in London and farther afield. We placed Dustboxes within the demonstrator gardens. We located Dustboxes with local residents. And we located Dustboxes at the official air-quality monitor adjacent to the Beech Street tunnel.[52]

Because the Beech Street tunnel has especially high levels of pollution in the form of nitrogen oxides and particulate matter, it was an area that local residents, the Museum of London, and the City of London were interested to gather additional data about. In the process of setting up Dustboxes alongside the Beech Street monitor, a City of London air-quality officer who provided access to the official monitor also undertook routine maintenance while at the site. Here, she collected a leftover particulate matter filter saturated with a black soot-like coating. The material residue of particles became evident as a layer of grime, which usually spreads through the air, circulates into the lungs and other organs of passersby, settles on urban structures, clogs instruments, and is taken up by more-than-human organisms.

The Dustboxes provided a way to compare pollution levels throughout the area while also looking at relationships across digital and vegetal sensors. However, the different modalities and durations of these sensing methods quickly became apparent, since moments of elevated air pollution did not correspond to

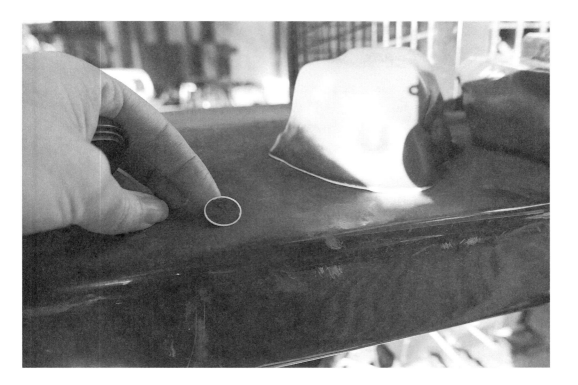

Figure 4.7. Installation of Dustboxes in the Beech Street regulatory air-quality monitor. This monitor, sited at the edge of the Beech Street tunnel and near the Barbican Tube Station, regularly registers high levels of nitrogen oxides and particulate matter. Photographs by Citizen Sense.

an automatic wilting of plants. Instead, vegetal sensors became a way to tune in to different approaches for cultivating less toxic atmospheric exchanges. In turn, the Dustboxes and other air-quality monitors throughout the area drew attention to where pollution was occurring and suggested how these patterns could inform proposals for developing other gardens in the area.

Distinct patterns of urban pollution began to surface through the Dustbox data and alongside the vegetal sensors. The demonstrator garden at the sheltered entrance to the Museum of London often showed the lowest pollution levels. However, when regional air pollution events occurred, this Dustbox showed levels similar to those of other monitors in the network. Pollution levels at the Beech Street tunnel location were consistently the highest, with pollution concentrations often exceeding the WHO twenty-four-hour guidelines. One resident on an upper floor of a Barbican tower typically recorded lower levels, likely due to the monitor height. Yet during windy conditions, this monitor would register some of the highest levels in the network. At all locations within this central area of London, however, levels of particulate matter were comparatively high overall, with monitors typically registering readings of 20 to 35 µg/m³ over a twenty-four-hour period. When winds circulated from the east (and from regional European locations), levels would often register at least 50 to 70 µg/m³, three to four times above the 2005 WHO guideline of 15 µg/m³ over twenty-four hours.

At the same time, the regulatory monitor in the Beech Street tunnel typically recorded levels of nitrogen dioxide at nearly double the EU twenty-four-hour standard, at 80 µg/m³. Looking more closely at pollution levels using our Airsift tool, we established that nitrogen dioxide was circulating from the nearby intersection and station to the west. At the same time, particulate matter was circulating from the east, both locally at the tunnel and regionally from farther afield. Digital sensors, vegetal sensors, and citizen sensors combined into observational techniques and experiencing entities registering both that pollution was occurring at high levels and that different approaches to addressing this problem might be created by reworking trajectories of sensing and action.[53]

How to Transform Citizen Sensing through Citizen Gardening

Citizen sensing and citizen gardening began to merge in the search for more breathable urban conditions. In this polluted location adjacent to the entrance of the Beech Street tunnel, the Dustbox and regulatory monitor shared the same airspace as one of the nineteen community-planted clean-air gardens. Existing concrete planters designed within the Brutalist architectural style of the Barbican were situated at the base of the forty-four-story Lauderdale Tower, where residents and volunteers had installed numerous air-quality plants. Small birch

(*Betula pendula*) trees occupied planters, where their waxy leaf surfaces were meant to trap particles, absorb nitrogen oxides, take up heavy metals, and phytoremediate the air and soil. Silverbush (*Convolvulus cneorum*), a plant with small leaves covered in numerous fine hairs that capture tiny particles, filled planters nearer to the entrance of Lauderdale Tower. The microclimate in this area was not only polluted but also dry and windy, making for conditions within which the more drought-tolerant silverbush was able to thrive.[54]

Citizen sensing could, in various ways, contribute to and align with citizen gardening and citizen infrastructure. But this juxtaposition of clean-air gardens, air-quality sensors, and highly polluted roadways inevitably raises the question of whether these relatively contained and diminutive planters would be able to mitigate the high levels of pollution occurring in this traffic-clogged area. Regulatory and citizen-led monitoring can establish that high levels of pollution occur and even demonstrate the location of emissions sources. However, the intensity and persistence of pollution often require further measures to address this problem. Air-quality gardens alone do not "solve" the problem of air pollution. Yet these engagements with more-than-human sensing also bring to the surface the dilemmas of how to realize feasible citizen actions with more intractable political processes. Such engagements work toward strategies for intervening within and transforming unbreathable urban conditions.

This inquiry into how to construct air-quality gardens investigates different practices for working with more-than-human sensors, spanning from identifying plants and outlining different planting scenarios for installing vegetation to building community networks. At the same time, these practices for working with vegetal environmental sensors and transforming urban air toward less hazardous and more collaborative environments point to the struggles that arise when attempting to address and mitigate air pollution. Citizen sensing and citizen gardening collide with different formations of power that can surface through practices of making and transforming environments.

From draining swamps to establishing plantations, cultivation is often an expression of power that materializes and reproduces dynamics of resource capture and extraction.[55] Planting is a way to reclaim, rework, and establish relations with land and more-than-human entities. Engagement with plants can become a form of resistance or otherwise inhabitation. As researchers and contributors to the Black/Land Project note, practices of urban gardening can form very different ways of engaging with land for Black gardeners, who also see this "as a way to stake a claim to permanency, education, economic citizenship, and community leadership, rather than only as a vehicle for food security."[56] Practices of gardening and inhabiting land can become forms of collective "geo-theorizing," where environments are made and remade as expressions of self-determination and openings into other ways of relating to land. In other words, this is a way of becoming citizens with land as well as worlds.[57]

In a similar register to the previous chapter, where people sought to protect green space to stake a claim to self-determination, the Black/Land Project describes how expressions of citizenship materialize through other and distinct ways of being in relation to and becoming with land. In parallel, Kimmerer discusses how gardening can generate practices of reciprocity that not only inform relations with land but also influence political relations and ways of undertaking exchanges with multiple entities. Reciprocity with the land is an invitation to reciprocity with all creatures. Such an orientation amplifies citizenships through the constitution of subjects, relations, and governance practices that work toward mutual exchange rather than domination.[58]

The cultivation of air-quality gardens signals the presence of air pollution in London, while proposing strategies for how to remake urban environments. Making and remaking environments are practices for attending to other entities, for cultivating their contributions to expand breathable worlds, while establishing and reinventing relations with more-than-human sensors. These are different ways of learning from and engaging with environments. They cultivate other relations that can depart from established inhabitations.[59]

The intervention that air-quality gardens make, both in the form of the demonstrator gardens we developed in collaboration with the Museum of London and in the form of the residents' nineteen clean-air gardens, is also worth considering in more depth in this discussion of how to construct air-quality gardens. While these gardens do not instantly absorb the many pollutants in London's air, they do constitute practices for transforming environments as multiple inhabitations, different durations, and collective exchanges that could generate more breathable worlds.

The Phyto-Sensor garden, events, and toolkits present scenarios for how to construct air-quality gardens and how to transform citizen sensing through multiple citizens' sensing and planting. These scenarios include how to work with community groups to start a garden project, how to identify sites for installing air-quality gardens, and how to select plants. They also encompass how to monitor pollutants, how to form different planting scenarios (from street trees to green walls), how to monitor along with vegetal sensors, and how to develop more expansive engagements with environments and more-than-human entities.

As collective environmental projects, these propositions for how to construct air-quality gardens resonate with community gardens. Yet even more, they connect with forms of near-future gardening. Many garden plots are now being developed not only as food-based projects but also as spaces for realizing environmental justice. These expanded forms of cultivation can especially be found in community gardens that address and mitigate climate-change inequality, where climate change and increasing urban temperatures make apparent the effects of redlining, lack of vegetation, and urban heat islands.[60] One such initiative, Groundwork USA, is working to make more "climate safe neighborhoods" in US cities by using digitized historical maps and data analysis to identify areas at risk of extreme heat and flooding due to climate change.[61] These analyses sense and overlay different forms of spatial inequality. Such groups are working toward more breathable urban environments by developing tree-planting initiatives and community gardens, which cultivate green space as a reconfiguration of environmental and political collectives.

Similarly, air-quality gardens could generate other counter-practices of cultivation that rework relations, entities, and environments through shared inhabitations with—and even beyond—pollution. Such gardens do not merely signal that an event has occurred or is occurring. Instead, they also materialize the relations at stake and environments that are harmed when pollution accumulates. The breathability of worlds multiplies through these exchanges that attempt to amplify rather than diminish environmental and social justice. Gardening becomes a sociopolitical practice for cultivating more breathable worlds along with more-than-humans, and a way to address and rework unprecedented environmental change through the co-constitution of citizens and worlds.

OPEN-AIR TOOLKITS

The open air has been a guiding concept and practice throughout this study. It designates ways of moving propositions into practice and experience. In this way, we explored how to put the Phyto-Sensor Toolkit to work in the open air.[62] We tested the toolkits through walking, workshopping, mapping, and inquiring into what conditions might be needed to construct air-quality gardens beyond the museum environment. These are ways of aerating gardens and toolkits. Yet this was less a project of reproducing air-quality gardens across London and farther afield than it was an attempt to explore how sensors, citizens, vegetation, and gardens come together in the open air. By testing the Phyto-Sensor Toolkit in these ways, we further inquired into different possibilities for activating more-than-human and multiple citizenships. Here, we encountered how pluralistic modes of sensing could open other environmental inhabitations.

As a thinker of multiple worlds, William James has suggested that by moving into the open air, experience might transform and undo the dogma of philosophy.

On one level, James uses the term "open air" to refer to something like "nature." Yet on another level, experience in the open air compels philosophy to undo a potentially self-referential focus to engage with worlds in process. Open air is not simply a synonym for the outdoors. Instead, it is an opening and an aeration of philosophy as well as subjects, experiences, and worlds. Because it is open-ended, the open air is propositional.

Breathability of practice can, then, generate breathability of thought. The open air instigates an aeration of theory and practice. Thought, as Crawley has similarly suggested, has been rendered airless and breathless through the pursuit of "categorical distinction," which has formed a rigid and racializing project.[63] By putting air into theory, Crawley proposes that difference and the "otherwise" as "a word that names plurality as its core operation" might open into forms of "radical sociality" that he calls "black pneuma."[64] Such a condition involves attending to "how breath moves and changes and performs in the world, the world that is made at the moment of the emergence of being together with others."[65]

Air—and the open air as I read it here through pragmatism, Black studies, Indigenous studies, and feminist technoscience—injects possibilities for otherwise and pluralistic thought and practice. In this way, the aeration of the Phyto-Sensor Toolkit involved working with museumgoers to investigate different sites and strategies for constructing and studying air-quality gardens within and beyond the Square Mile. Concrete proposals and observations, as well as different conditions for being and becoming citizens, assemble through these open-air excursions. They form other conditions of exchange, reciprocity, and searching toward more breathable worlds. So too, multiple citizens materialize through these co-constitutive and expanded relations across entities and experiences.

How to Aerate a Phyto-Sensor Toolkit

Because the construction of the demonstrator air-quality gardens took place primarily as an institutional collaboration, Citizen Sense hosted a workshop and walk at the Museum of London where the Phyto-Sensor Toolkit could be opened into dialogue with interested museumgoers. The workshop and walk allowed for the further development of the toolkit by drawing on input from participants. While existing residents had made the above-described monitoring projects (and were involved in monitoring with Dustboxes and meeting to describe their monitoring activities), the publics that attended the museum were much more diffuse, with visitors to the *City Now City Future* exhibition traveling from locations across London, the UK, and the world.

Citizen Sense developed the Phyto-Sensor Toolkit as a resource and guide to aid in the citizen-led research and development of air-quality gardens. The

Figure 4.8. Air-quality garden walking tour with visits to air-quality plants, including *Euphorbia* and lamb's ear. Photographs by Citizen Sense.

toolkit focuses on how phyto-sensing processes can bioremediate air quality, especially in urban settings. As a pedagogical toolkit, it introduces participants to an array of herbaceous and woody plants, as well as trees, that respond to and mitigate pollution. The toolkit allows people to identify plants in the field while at clean-air garden and green infrastructure sites and to learn more about how they bioindicate or mitigate pollution. Consisting of ten sections, the toolkit first provides an introduction to phyto-sensing and an explanation of citizen sensing. It then gives an overview of the problem of air pollution and describes the Low Emission Neighbourhood initiative and garden. The core content of the toolkit includes a series of plant-identification cards that describe key vegetal responses to air, soil, and water pollution. Also included in the draft version of the toolkit is an outline of the workshop and walk (where the provisional version of the toolkit asked for participants to record comments and propositions for the final toolkit). As part of the walk itinerary, the toolkit introduces the air-quality gardens and air-quality monitoring infrastructure in the Square Mile of London. The final sections of the guide include suggested planting scenarios for constructing air-quality gardens in relation to likely emissions sources, and a list of resources for further researching air-quality plants and gardens.

To test the Phyto-Sensor Toolkit, we held a workshop and walk as part of the Barbican OpenFest, an annual free event to explore the outdoor and indoor spaces around the Barbican. On the day of the walk, a Saturday in mid-March when temperatures might ordinarily be mild, we were instead visited by a cold snap dubbed the "mini-Beast from the East." Grappling with temperatures of –7 degrees Celsius with wind chill, we worried that people might not attend or be equipped to handle the temperatures that were unusually cold for London. However, many participants did turn up sporting full-body outerwear suitable for extreme winter sports, as well as wearing heavy gloves, woolly scarves, and bobble hats, fully layered up and ready for an investigation of air-quality gardens.

After a brief introduction to the Phyto-Sensor project and an overview of the workshop, walk, and toolkit, we set out to look at air-quality plants in the area. We started our walk at the two demonstrator air-quality gardens at the entrance to the Museum of London. Here, we discussed the different plants installed, their bioindicating and mitigating properties, and how they process pollution in different contexts and in relation to other organisms. Paul McGann of Grow Elephant joined Citizen Sense for the workshop and walk, during which he explained the development of the planters and the installation and care of the plants.

With the frigid temperatures, plants were not yet in full bloom, but many were still identifiable when compared to the field guide in the Phyto-Sensor Toolkit. We struck a balance between lingering to investigate the fine or hairy leaves of

Figure 4.9. Air-quality garden walking tour across the Barbican site. Photographs by Citizen Sense.

plants that capture particles, to walking briskly to the next stop where we might be able to fend off the cold. We walked around the Barbican podium, discussed residents' initiatives to address air quality, and made our way to the Beech Street monitoring station. As mentioned previously, this enclosed hard space of concrete and traffic tends to have especially high pollution levels. We then looked at how urban architecture, vegetation, monitoring, and urban processes assemble to create more or less polluted environments, and discussed other approaches we could take to construct more breathable conditions.

In the open air, the Phyto-Sensor Toolkit began to unfold as a logbook, plant-identification guide, map, and observational device for recording, considering, and proposing different ways of engaging with vegetal sensors and air-quality gardens. Given the cold temperatures, this unfolding was at times marked by a rapid jog-walk to the next garden. At various stops, we would wave to a stand of *Euphorbia* and quickly move on to observe lavender, lamb's ear, and jasmine while hopping in place to learn about clean-air gardens developed by residents as part of the City in Bloom initiative in 2017. We also discussed the proliferation of air-quality plants and gardens in numerous other locations that we would not be visiting, including at the Barbican Tube Station, where a community garden of air-quality plants had been installed to green the urban canyon of the station. On a disused rail track, coral bells and ivy filled planters, where their broad leaves and masses of green surfaces were meant to capture particles and nitrogen oxides. We speculated about how large a planted area would need to be, especially in this area of London, to take up pollutants in an effective way. One participant asked if there was a way to measure plant mass needed in relation to pollution levels. Many people wondered: Was this a feasible strategy, given the intensity and pervasiveness of air pollution in London? Did we need a how-to guide for aerating London's atmospheres on a wider scale?

As we expeditiously made our way to the Moor Lane pop-up garden, we discussed how air-quality gardens could also be a way to make more evident the role and benefit of vegetation in improving air quality. The Moor Lane garden had developed through a considerable volunteer effort. Populated with corrugated steel pipes that sprouted a burst of air-quality plants, the site provided a striking visual display that collided with the hard concrete and brick edge of the Barbican podium and parking garage. Numerous air-quality plants jostled together in this space, including ivy, ferns, birch, *Hebes*, silverbush, and juneberry. Plants with smooth but small leaves such as *Hebes*, as well as vigorously spreading plants like ivy, came together in a display of plants that we discussed and imagined to be sensing, absorbing, capturing, and filtering air pollution. We noted that small trees such as juneberry can mitigate pollutants including particulate matter and

Figure 4.10. Air-quality garden walking tour visits the pop-up garden developed through the City in Bloom initiative. Photographs by Citizen Sense.

Figure 4.11. Barbers' Physic Garden at the edge of the Barbican and Museum of London sites; viewing plants in the Barbers' Physic Garden, including *Camellia*, previously used for treating asthma. Photographs by Citizen Sense.

nitrogen dioxide as long as their canopies do not mass together and trap pollution at street level. At the same time, this species is particularly beneficial for
wildlife, providing berries, pollen, and nectar for birds and insects. The cultivation of urban environments began to draw in multiple other organisms that form
ecologies specific to air-quality gardens.

We then made our way to the final stop along our walk, visiting the Barbers'
Physic Garden to consider the restorative and curative qualities of plants. This
historic garden was developed in the sixteenth century and was a site to which the
botanist John Gerard later contributed.[66] The garden was originally developed
with plants required by surgeons, and the most recent version of the garden was
planted in 1987 on a derelict bomb site. In this series of plots, forty-five different
species of herbs lined gravel pathways, with each specimen having a placard
noting its distinct properties for treating wounds, bruises, and burns. In this site
of former bomb damage, strewn with the ruins of a Roman fort and Victorian
walls, we investigated the relationship between plants and health. We considered
how plants as curative organisms could contribute to healing environmental health
problems. Finally, we asked: Could we make a twenty-first-century how-to guide
and Physic Garden for restoring the health of environments?

Once we had returned to the Museum of London, we regrouped over a much
needed tea-and-biscuit break before reviewing the air-quality garden walk and
beginning a discussion of the Phyto-Sensor Toolkit. Drawing on observations
and propositions that participants recorded along the way, as well as generating
additional proposals that arose in the course of the workshop, we spent some
time thinking and working through what communities would need to construct
air-quality gardens in areas of London and farther afield. Paul offered advice on
plants and gardening techniques, while we reviewed the difficulties of working
with local councils and the complexities of ensuring maintenance of vegetation
so that air-quality gardens would not expire from neglect.

The final version of the Phyto-Sensor Toolkit incorporated these proposals
and observations from workshop and walk participants. As a how-to guide for
air-quality gardens, the toolkit developed through several iterations in the open
air. The construction of gardens, the installation of vegetal and digital sensors,
the testing and expanding of the instructions and resources through a workshop
and walk, and the compiling of the toolkit as a print and online manual available
on the Museum of London website all became ways of aerating the toolkit and
expanding the designations and practices of citizens and sensing. We developed
the toolkit through these open-air modes of inquiry so that it could contribute to
additional air-quality gardens and community projects. Just after we published
the toolkit online in May 2018, we learned that community groups were taking

up the toolkit and using it as a guide for making air-quality gardens, including on Clean Air Day. Aeration is, then, an invitation to incorporate, attend to, and participate in more-than-human sociality and sensing. It presents an openness to expanded and more-than-human modes of sociality that work toward more breathable worlds.

How to Multiply Citizens in Worlds

Plants are participants in urban ecologies, and they contribute to urban environmental communities. The Phyto-Sensor Toolkit tunes in to these expanded processes of environmental sensing that occur across multiple entities. This is another way of thinking about and working with sensing beyond an anthropocentric approach to environments, where organisms and environmental relations are differently brought into view and activated through their interrelations with and experiences of pollution.

As Crawley suggests, "everyone is held within breathing as process."[67] And yet, this breathing involves multiple engagements that are differently situated and experienced. Humans unequally experience exposure to air pollution, which can in turn even affect their capacity to sense. Plants absorb and exchange pollutants, while multiple other organisms are caught up in sensing and processing the effects of polluted atmospheres. Multiple organisms act as sentinels, bioindicators, and more-than-human sensors that express the conditions of environmental toxicity and change. These practices constitute differential and at times collaborative modes of sensing. Such expanded sensing practices offer different ways of cultivating urban air and working toward more breathable worlds by tuning in to the unevenly shared conditions of breathing, sensing, and exchanging atmospheres.

These collective contributions to environmental sensing provide an invitation to cultivate more expansive citizens and worlds in the making as "pluralistic realism."[68] As discussed throughout this study, citizen-sensing practices often challenge the "one-world world" of scientific realism. Different experiences, modes of evidence, and encounters with air pollution materialize through these practices, which do not assemble into a singular epistemological or ontological plane.[69] So too does more-than-human and vegetal sensing shift and multiply the worlds of experience that could be brought to bear on the problem of pollution.

Such an approach requires that other worlds of relevance register as significant, rather than attempting to fit these practices into a singular world of experts and amateurs, humans and more-than-humans. Citizen sensing can be an example of a project that is meant to aid an already established trajectory of scientific inquiry.[70] Yet this chapter outlines an approach to multiple citizens to indicate how worlds, political subjects, relations, and actions form through different engagements with sensing organisms and sensing technologies. As John Law notes, these are the conditions that challenge a "one-world world," since practices are a way of activating different realities. Such practices have sociopolitical relevance, since they draw together different collectives, activate different versions of the real, and require contingent and specific strategies for working with and across pluralistic conditions.

Air-quality gardens and more-than-human sensing practices are conditions that might be encountered and cultivated in the middle of things. Pollution and environmental change are unequally bearing down on humans, more-than-humans, and ecologies. Yet these are also conditions that generate different political subjects and political work. Such conditions require strategies for being and becoming citizens of worlds along with multiple other entities in the open air. This turn toward the open air involves attending to concreteness and worlds of experience.[71] By working from within the particulars of multiple and differential experiences, it is possible to realize the possibilities of pluralism as what James calls "a translocation of experiences from one world to another."[72] As discussed here, this translocation of experience spans from the vegetal to the digital and beyond. Worlds are not a total surround. Instead, they are vectors of engagement and experience. They form a multiverse or pluriverse, rather than a universe.[73] Worlds involve constitutive processes, exchanges, and intensities. They establish conditions of breathability and reciprocity, as well as of unbreathability and subjugation. Environmental sensing unfolds through these distinct yet pluralistic registers to explore and establish how multiple modes of sensing make and remake citizens and worlds.

HOW TO CULTIVATE RELATIONS

Environmental sensing across multiple entities requires and contributes to the reconstitution of citizens and worlds. In this pragmatist orientation, citizens and worlds are co-constituted and in the making, as well as the unmaking. Multiple citizens involve not merely the multiplication of worlds but also the expansion of ways of being citizens, engaging in relations, and sensing, exchanging, and breathing together. The hard lines between humans and other organisms could begin to blur and be reworked as a vegetative process in this approach to multiple citizens. As Monique Allewaert suggests, drawing on eighteenth-century investigations into coloniality and botany, this could be a way of "vegetating" the subject through and toward more ecological configurations (and even against the citizen of nationalism and colonialism) that complicate "clear distinctions

Figure 4.12. Phyto-Sensor Toolkit workshop, participants developing proposals for air-quality gardens to be included in the toolkit. The final version of the toolkit was made available through the Museum of London and Citizen Sense websites. Photographs by Citizen Sense.

between human beings and the natural world."[74] Through a discussion of *marronage*, Allewaert identifies how ecological configurations of subjects and environments could generate political force by reworking the categories of human and nature while recasting and abandoning categories of slaves as objects. Practices of making and remaking citizens and worlds toward expanded ecologies involve political work, as discussed earlier, that has distinct consequences for the breathability of worlds.

Within processes of constructing, testing, and circulating plans for air quality, multiple citizens surface as such vegetating entities. They are more-than-human collaborators, plucky cultivators, enduring sensors, open-air exchangers, aerating world makers, and political workers. This aeration of citizens and citizenship provides a guide for how to reconstitute citizens through relation, action, and environmental inhabitation. This is an aeration and ventilation of citizens, which seeks to cultivate multiple and recomposed subjects, relations, and worlds. Here, pluralism is not simply an accumulation of more entities. Instead, it is a site of possibility and a provocation to constitute citizens through relation and action. This investigation into multiple citizens is a proposition to consider how worlds are composed through other experiences, and how this might in turn generate forms of "engaged pluralism."[75] In working toward pluralism, James sought to move away from foundationalism, substantialism, and fixed terms, and toward worlds of experience and their a/effects. Yet these forms of action often lead to struggle when attempting to make sense of and compose breathable worlds across different experiences.

As *Citizens of Worlds* emphasizes, encounters with difference and struggle are persistent features of political engagement and democratic life. These conditions are political, moreover, because the conditions of breathability cannot be taken for granted. Air pollution is an often intractable problem that generates differential conditions of struggle. Multiple practices proliferate in attempts to aerate the polluted conditions of urban environments while transforming the confined domains of extraction and despoliation, the diminished breadth of violence and extirpation, and the airless spaces of thought and politics. These are struggles for environments, for constitutive exchanges, for action, and for reciprocity. "A struggle to achieve reciprocity," as Kimmerer suggests, involves working toward breathability and democracy of and for all species in ways that could reinvent possibilities for political subjects, justice, and governance.[76] In this guide for how to construct air-quality gardens and vegetate citizens, you might find also suggestions for how to cultivate and work toward reciprocity by transforming the subjects and relations through which environmental problems come to matter and are acted upon.

AIRKIT TOOLKIT

AIRKIT

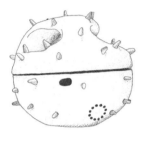

The AirKit toolkit creates a citizen-sensing infrastructure for monitoring air pollution. It consists of a logbook, a second-generation Dustbox, an Airsift platform, and a tool for writing data stories. More information is available at https://citizensense.net/projects/#airkit. Illustration above by Sarah Garcin, illustration below by Andrea Rinaldi; courtesy of Citizen Sense. This toolkit can be found in a more extensive form online at https://manifold.umn.edu/projects/citizens-of-worlds/resource-collection/citizens-of-worlds-toolkits/resource/airkit-logbook.

Conclusion

SENSING CITIZENS

How to Collectivize Experience

What we were discussing was the idea of a world growing not integrally but piecemeal by the contributions of its several parts. Take the hypothesis seriously and as a live one. Suppose that the world's author put the case to you before creation, saying: "I am going to make a world not certain to be saved, a world the perfection of which shall be conditional merely, the condition being that each several agent does its own 'level best.' I offer you the chance of taking part in such a world. Its safety, you see, is unwarranted. It is a real adventure, with real danger, yet it may win through. It is a social scheme of co-operative work genuinely to be done. Will you join the procession? Will you trust yourself and trust the other agents enough to face the risk?"

William James, *Pragmatism*

Throughout this study, I have attended to how sensing technologies activate the "citizen" as a political entity. I have considered how these political subjects variously converge and diverge in struggles to monitor environments. Multiple modes of citizenship circulate throughout these struggles, including the atmospheric, the instrumental, the speculative, the data-oriented, the multiple, and the sensing. Yet many more citizens surface here, from breathing citizens to contradictory citizens, digital citizens, good citizens, insurgent citizens, neoliberal citizens, planetary citizens, and scientific citizens. Indeed, many of these citizens proliferate in each chapter when discussing how citizen sensing is put to work.

Rather than encountering these citizens as variations of human subjects, however, this research offers a more pluralistic proposal for *citizens of worlds*. This concept attends to how subjects and worlds are co-constituted. By monitoring environmental problems such as air pollution, citizens operate as distributed and collective sensors that tune in to and act upon environmental problems. Polluted environments also make demands on how subjects experience, respond to, and act on these worlds. Political subjects and collectives form along with these worlds of experience.[1] Citizens and worlds are in the making.

In the context of this discussion on environmental sensing and air pollution, I especially consider how citizens of worlds signal conditions of breathability. The breathability under discussion here especially pertains to air pollution, as a residue from the ongoing extraction and consumption of fossil fuels. The construction of pipelines that began this study, as well as the fracked gas, the incessant traffic, and the multiple other polluting industries and activities discussed here, contribute to toxic air and a warming planet. As one of the most significant environmental and health problems globally, air pollution contributes to nearly nine million deaths every year, with many more people living with chronic health problems due to air pollution. It also contributes to clogged atmospheres that affect multiple more-than-human organisms.

While air is often described as something held in common, and breathing is frequently described as a universal condition of human life, the actual constrictions and differential compositions of atmospheres can be elided or overlooked in gestures toward universality. Struggles for breathability indicate how air and air pollution are unevenly distributed. Toxic atmospheres are unequally experienced and endured. Bodies are diversely affected by the conditions in which they endure or expire. Such atmospheres can contribute to practices of "combat breathing," a concept from Frantz Fanon that has informed numerous writers who work through the unbreathability of social inequality, environmental injustice, racism, colonialism, and contradictory citizenships to propose ways of breathing otherwise.[2]

This discussion of breathability is not only a reference to the literal (or even metaphoric) quality and condition of air and atmospheres; even more, it is an analysis of social, political, environmental, and more-than-human exchanges that make citizens and worlds. Such a concept orients inquiry toward how citizens and worlds form, gather strength, and dissipate through ongoing struggles for breathability. Although breathing is a collective endeavor, it is differentially experienced.[3] Breathing here is not simply a routine inhalation and exhalation of air. To suggest otherwise runs the risk of reverting to a "one-world world" that would overlook the struggles of those who are fighting for more breathable worlds. Industries are entitled to pollute the air, wealthy people have greater privilege to directly or indirectly emit and avoid damaging pollutants, and poorer people bear the injustices of living next to polluting industries and transport infrastructure. Breathing is not guaranteed for all, and possibilities to be and become citizens vary significantly within more or less expansive or constricting worlds.[4]

In this way, the uneven distribution and experience of air pollution can highlight environmental injustices in the sacrifice zones of rural industrial areas, urban areas experiencing poverty and gentrification, and financial centers where the exchanges of communities and more-than-humans can diminish under the weight of ongoing urban development. To discuss these diverse conditions together is

Figure C.1. Dalmain Primary School green screen and air-quality sign; Dustbox installed as part of the AirKit project in Forest Hill, London, 2020. Photographs by Citizen Sense.

not to flatten them into a single record of air pollution and its impacts. Instead, such an approach demonstrates how these worlds form through struggles and differential registers of experience.

To grapple with the lived experiences of these struggles, this study has developed the concept and practice of open-air instrumentalisms. Open-air instrumentalisms orient toward the how-to as a mode of collective inquiry, which further forms an organizing thread throughout this text. This inquiry asks: How is it possible to sense the air? How do different political encounters and relations struggle against environmental injustice? What citizens and citizenships materialize here? And how might breathable worlds be made? The toolkit forms an initial invitation for how to operationalize citizen sensing through these modes of inquiry. The toolkit intervals that punctuate this text document how we assembled various sensors and components to test and query.

Yet as *Citizens of Worlds* demonstrates, such instructional approaches quickly become open-ended practices for how to make worlds, and how to make worlds matter. They do not, as James suggests in the epigraph to this conclusion, lead to certain outcomes. But they could lead to more reciprocal ways of working together, as crucial components of democratic exchange and breathability. By moving into the open air, citizen-sensing practices are activated as they encounter environmental problems.[5] In the process, the citizens and citizenships that materialize are neither as singular nor as singularly effective as citizen-sensing scripts might ordinarily propose. Instead, different registers of political engagement surface through these struggles for more breathable worlds.

The three case studies that I discuss in this book describe how communities, Citizen Sense researchers, and many other collaborators have found ways to cooperate and undertake collective inquiry. We were, in different ways, sensing citizens who attempted to work through experiences of air pollution, environmental sensors, digital platforms, citizen data, planning documents, governance structures, community meetings, atmospheric science, health conditions, and much more. We researched the multiple ways people monitor and observe environmental problems, including how to collect and analyze data on air pollution and emissions sources, how to mobilize observations, how to lobby industry and governments to take action, and how to engage in collaborations and protests as different worlds in the making collide with inequalities and power imbalances.

The "citizen" in "citizen sensing" might seem to confer certain political and technical skills abstractly and evenly on all users of these technologies. While exclusions and inequalities might ordinarily pertain to the nation-state as the site of (differential) belonging and privilege, in the realm of technological citizenship these conditions might have more to do with "capacities" of citizens, not necessarily as a form of national belonging or grouping, but more as an ability to contribute. But this is potentially a one-world world way of understanding citizenship. It suggests there is one correct register of capacity and contribution where certain subjects can perform in these registers. These conditions could determine who can and cannot be or become a citizen within a universal technopolitical milieu. As the NoDAPL protectors and pilots discussed in the Introduction demonstrate, capacity can involve pushing technology to its limits to appropriate and invert the usual means of using it, especially when attempting to materialize lived experiences and worlds. These capacities might not be recognized as legitimate or even legal forms of participation. However, such practices can generate distinct political and technological engagements that challenge environmental destruction and injustice.

While technologists are busily appending "citizen" to digital devices, these technologies turn out to be neither so clearly open nor participatory. Instead, they can generate power imbalances of all sorts.[6] The citizen is often an assumed figure within citizen science, citizen sensing, and related forms of public engagement, which mobilize the term "citizen" to suggest that a practice or technology might be equally available to everyone, or even spark democratic modes of participation. However, very different political engagements can form along with these devices. Even those "citizens" who follow the instructions and generate legible evidentiary outputs might find that their practices and findings are not recognized. Digital sensing technologies and the citizenship advantages they are meant to confer could reproduce or amplify existing struggles over citizenship and legitimacy. Devices shift and transform with the environmental problems people attempt to address within and through circuits of politics and power. In this sense, the citizenly promise of sensors does not equally or so readily materialize as instant guarantors of engagement.

Environmental sensing toolkits could activate particular technological expressions of citizenship and articulate distinct affiliations and exchanges that support the formation of certain political subjects and not others. However, *Citizens of Worlds* seeks to redirect the focus on citizenship less toward the agency of a singular human subject, or the more neutral registers of digital participation. Instead, it orients more toward the distributions of effect and effectiveness formed in collective struggles for breathable worlds. Such a focus attends to the relations, processes, multiple entities, environments, politics, and ontologies that activate and sustain citizens and worlds. These citizens and worlds come into being, proliferate, subside, and are erased. Because they are in the making, they do not fit neatly into an official genealogy of citizenship. In this sense, the struggle for breathable worlds is often undertaken by unauthorized participants, those whose struggle to breathe is a struggle not just to have their voices heard but also to make their worlds matter.

This citizens-of-worlds approach to citizen sensing deliberately engages with these uneven political, technical, and social registers of environmental monitoring. It considers not a limited range of legitimate or capable actors, practices, and devices, but instead attends to pluralistic ways of engaging with, observing, witnessing, documenting, communicating, and acting on environmental problems. Here citizens are less stable and universal individuals with settled rights and responsibilities and more flickering figures that shape-shift through changing relations. Rather than neatly bounding the citizen, it might be possible to unravel the loose threads of citizenship to demonstrate how it sprawls into an unruly composition that is more characteristic of the goings-on of democratic life.

As a concept, "citizens of worlds" seeks to create an approach to citizenship as a register of collective experience. This approach looks at the concrete articulations of citizens and citizenship that occur, and how these variously challenge and remake political subjects and engagement toward more breathable conditions. Instead of singularly defining *the* citizen, such an approach travels with the conditions, communities, and feelings that settle into citizenly engagements. By attending to the struggles across different citizenship practices, I have sought to trace out and inhabit the open-ended and restrictive forms of citizenship mobilized through citizen sensing as an emerging form of political engagement with environmental problems. At the same time, when taking up a project of attending to proliferations, it is always necessary to address how some forms of citizenship proliferate to the detriment of others.[7]

Sensing citizens form not just as they attempt to make sense of worlds under threat and suffering from environmental destruction. They also materialize through practices that attempt to operationalize sensing toolkits to generate less extractive relations—with technology, environments, and other entities. Technology is more than an artifact. It extends to and encompasses the entities, relations, and expanded milieus that inform how technics materialize and unfold. This means that citizen-sensing devices do not automatically generate intended political effects. Instead, they are put to work in ways that co-constitute citizens, worlds, and breathability in searching after political possibility. By seeing technologies as extended relations and fields of influence, it could be possible to create more constructive technological engagements that attend to pluralistic citizens, relations, milieus, sociality, and worlds in the making. Unlike in James's thought experiment, there is not one author of one world that materializes here. However, worlds involve piecemeal projects that arise through multiple contributions and experiences. The commitment to build worlds through cooperative work and resonant experiences is the generative spark that binds them together. Sense-making then becomes critical to the doing and sustaining of democratic worlds.[8] This is how you might collectivize experience.

Figure C.2. Dustboxes, a particulate-matter sensor for monitoring air quality developed by Citizen Sense. Photograph by Citizen Sense.

Notes

INTRODUCTION

1. Chariton, "Drone Pilots Exposing Oil Police Violence." For more extensive studies of Indigenous resistance and environmental (in)justice at Standing Rock, see Estes, *Our History Is the Future*; and Whyte, "The Dakota Access Pipeline."

2. See also Drone2bewild (interviewed by Paulette Moore), "Drone Footage of Dakota Access Pipeline"; Real News Network, "Police Are Shooting Down Aerial Drones." For more extensive analysis of protest as well as environmental monitoring with drones, see Kaplan, "Atmospheric Politics"; Elliott et al., "Drone Use for Environmental Research"; and Fish, "Crash Theory."

3. Duarte, *Network Sovereignty*. For a discussion of computing otherwise, see Pritchard, "The Animal Hacker"; Amrute and Murillo, "Introduction"; and Philip, Irani, and Dourish, "Postcolonial Computing."

4. Chariton, "Drone Pilots Exposing Oil Police Violence." For a more extensive discussion of Indigenous science and environmental change, see Whyte, "Indigenous Science (Fiction)."

5. See Clark, *The Poisoned City*; Hanna-Attisha, *What the Eyes Don't See*; Fennell, "Are We All Flint?"; and Pulido, "Flint, Environmental Racism, and Racial Capitalism."

6. For a more extensive discussion of these dynamics, see Hemmi and Graham, "Hacker Science versus Closed Science"; Kimura, "Citizen Science in Post-Fukushima Japan"; Kenens et al., "Science by, with and for Citizens"; and Plantin, "The Politics of Mapping Platforms."

7. See Global Forest Watch.

8. For a sense of the range of community and citizen science projects monitoring air quality, see the Air Sensors International Conference 2018 program archive, which includes slides and videos of project presentations. For a more extensive discussion of one of these projects, IVAN (Identifying Violations Affecting Neighborhoods), see English, Richardson, and Garzón-Galvis, "From Crowdsourcing to Extreme Citizen Science."

9. Gabrys, "Planetary Health in Practice."

10. For a discussion of the connection between wildfires, air pollution, and sensing, see Gabrys, "Sensing a Planet in Crisis."

11. This concept of world making travels across multiple texts that I engage with throughout this book. While numerous researchers take up this term in different and diverging ways, I especially situate this discussion in relation to William James's discussion of world making in *A Pluralistic Universe*.

12. James, *Pragmatism and Other Writings*, 27. While I draw on James's notion of the open air, the general sensibility to test and remake concepts in relation to lived experience runs through a number of pragmatist texts.

13. Alfred North Whitehead did not use the term "open air" as such. Still, James's writings influenced his commitment to pluralism, which Whitehead described in his own work as "pluralistic realism." In this way, his emphasis on contingency and experience resonates with James's focus on worlds in process. See Whitehead's *Process and Reality*, 78, 88.

14. I have previously briefly written about "atmospheric citizenship" in the context of climate change and managing the global commons of the atmosphere as articulated through creative-practice projects. In other writing I have explored how wireless and digital technologies operate through atmospheric modalities to form atmospheric media. This current work on citizen sensing more directly engages with the atmospheric registers of air pollution, breathing, and citizenship while also drawing on this earlier work. See Yusoff and Gabrys, "Climate Change and the Imagination"; and Gabrys, "Atmospheres of Communication." In addition, as I discuss throughout *Citizens of Worlds*, many more writings at the juncture of atmospheres and air pollution inform this study of atmospheric citizenship, including Alaimo, *Bodily Natures*; Choy, *Ecologies of Comparison*; Fortun, "From Latour to Late Industrialism"; Graham, "Life Support"; and Mark Whitehead, *State Science and the Skies*. Writing recently in a more biopolitical rather than pluralistic register, Asher Ghertner suggests how the "airpocalypses" in Delhi, India, generate atmospheric citizenships that require governing "the citizen body" as a distributed composition. See Ghertner, "Airpocalypse."

15. There is an extensive and expanding range of work on digital (and data) citizenships that this book does not have space to survey. While throughout I draw on my earlier analyses of citizens and digital technologies, especially as developed in Gabrys, "Programming Environments," I am also in conversation with ongoing research in this area, including Hintz, Dencik, and Wahl-Jorgensen, *Digital Citizenship in a Datafied Society*; Isin and Ruppert, *Being Digital Citizens*; and Powell, *Undoing Optimization*. Also in the background are studies on citizens and smart cities—see Shelton and Lodato, "Actually Existing Smart Citizens"; and Datta, "The Digital Turn in Postcolonial Urbanism."

16. Citizen science as a field and topic of research has been proliferating over many decades now, along with civic science, public engagement, participation, and many more related areas of theory and practice. While this study does not undertake a comprehensive survey of these literatures, several texts informing this research include Irwin, *Citizen Science*; Michael, "Publics Performing Publics"; Lave, "The Future of Environmental Expertise"; Chilvers and Kearnes, *Remaking Participation*; Felt and Fochler, "The Bottom-up Meanings"; Waterton and Tsouvalis, "An Experiment with Intensities"; and Jasanoff, "Technologies of Humility."

17. Gabrys, "Citizen Sensing, Environmental Monitoring."

18. Goodchild, "Citizens as Sensors." While this text is often referenced as a sort of originary use of "citizen sensing," Goodchild instead referred to "citizens as sensors"

who might crowdsource map data within the context of volunteered geographic information (VGI). There is actually very little by way of sensor technologies—or sensing as a
practice—in Goodchild's text. However, this is an early and interesting example of participatory mapping, as developed by many researchers often working in geography. For
parallel examples, see Elwood, "Volunteered Geographic Information"; and Zook et al.,
"Volunteered Geographic Information." Earlier uses of "citizen sensing" in reference
to environmental sensing technology by potentially grassroots groups of participants
seem to be within the HCI community and tech companies. For instance, the Center for
Embedded Network Sensing (CENS) used the term "participatory sensing" in Burke et al.,
"Participatory Sensing." However, here I want to complicate this genealogy and usage
further by attending to the fluidity of a practice that has not yet settled into form, and by
engaging with how the scripts of technology companies do not lead to a direct unfolding
of devices in the world toward certain outcomes. In this way, while this study also engages
with citizen science and aligned practices, it does not adopt a taxonomic or classificatory
approach to these practices, which are often further presented as a hierarchical ordering
of modes of participation, perhaps under the influence of Arnstein's "ladder of participation." See Arnstein, "A Ladder of Citizen Participation."

19. More information on the Citizen Sense project is available at http://citizensense
.net.

20. The statement attributed to Theresa May is: "Today, too many people in positions
of power behave as though they have more in common with international elites than with
the people down the road, the people they employ, the people they pass on the street. . . .
But if you believe you are a citizen of the world, you are a citizen of nowhere. You don't
understand what citizenship means." There was significant backlash after this statement,
with many people adopting the label "citizen of nowhere" on social-media profiles and
adding it as graffiti in London spaces. See Bearak, "Theresa May."

21. As Pheng Cheah suggests, "cosmopolitanism is about viewing oneself as part of
a world, a circle of political belonging that transcends the limited ties of kinship and
country to embrace the whole of deterritorialized humanity." Cheah, *What Is a World?*, 3.

22. See Kleingeld, *Kant and Cosmopolitanism*, for a discussion of these permutations
of world citizenship and cosmopolitanism, especially in relation to Kant. As she suggests,
in a Kantian reading, a citizen of the world is not necessarily someone who is unfixed
and rudderless, but instead can be someone who is rooted in place and yet has a concern
and consideration for world affairs, and is not limited by the nation. However, this project may involve caring for a very particular kind of world (formed through European
colonialism), and by a particular kind of white male European citizen–subject.

23. In discussing the work of Karl Jaspers, Hannah Arendt suggests, "The establish-
ment of one sovereign world state, far from being the prerequisite for world citizenship, would be the end of all citizenship. It would not be the climax of world politics, but
quite literally its end." Arendt, "Karl Jaspers," 82. This particular configuration of politics
as plurality can be found in the work of Carl Schmitt, who writes, "The political world
is a pluriverse, not a universe." Schmitt, *The Concept of the Political*, quoted in de la
Cadena, "Indigenous Cosmopolitics in the Andes," 341. De la Cadena queries this formation of politics by considering the other-than-humans that could also be constitutive of
the political.

24. Law, "What's Wrong with a One-World World?" See also de la Cadena and Blaser, *A World of Many Worlds*. In a different but resonant register, Kathryn Yusoff refers to the "global-world-space" where the world is a singular entity available for conquest, which in turn informs relations across subjects and environments. See Yusoff, *A Billion Black Anthropocenes or None*.

25. For instance, Balibar notes that the emergence of the citizen of the world was in relation to a European colonial world that enabled specific transnational flows of commodities and people. See Balibar, *Citizenship*, 71. See also Lowe, *The Intimacies of Four Continents*.

26. These more-than-state formations of citizenship are a common refrain within citizenship studies, and can be found in work ranging from Balibar, *Citizenship*, to Braidotti, *Transpositions*. In a different way, Indigenous and Black studies literatures discuss refusing the nation—and its forms of citizenship—as units of belonging because of the exclusions, diminutions, violence, and dispossessions that have been and are undertaken under the label of the citizen. See A. Simpson, *Mohawk Interruptus*; and Sharpe, *In the Wake*.

27. It bears mentioning that "constitution" is a key way in which Balibar theorizes the formations of citizens and citizenship. Coincidentally, "co-constitution" has become a word commonly used to refer to the (participatory) formation of entities and practices, for instance, within collaborative research projects. These multiple versions of constitution are at play here. See Balibar, *Citizenship*.

28. In addition to digital citizenship literature noted earlier, there are now numerous works that challenge the easy or singular promises of greater participation (or even disconnection) that digital technologies and makerly activities would bestow. For example, see Irani, "Hackathons"; Kelty et al. "Seven Dimensions"; and Bucher, "Nothing to Disconnect From?"

29. As Isin further writes, citizenship is a conception of "dominant" groups that produce the alterities of citizenship. Isin, *Being Political*, 5.

30. Writing about the *noncitizen*, Haraway has similarly suggested, "the discursive tie between the colonized, the enslaved, the noncitizen, and the animal—all reduced to type, all Others to rational man, and all essential to his bright constitution—is at the heart of racism and flourishes, lethally, in the entrails of humanism." Haraway, *When Species Meet*, 18.

31. Berlant makes this point and further notes, "Citizenship is the practical site of a theoretical existence, in that it allows for the reproduction of a variety of kinds of law in everyday life." Berlant, "Citizenship," 38–40.

32. How citizenship can be restricted or limited, especially in relation to race (as well as class, gender, and other categories of the social), is discussed more extensively by Nelson in *Body and Soul* in reference to national as well as economic, biological, and medical forms of citizenship. Nelson's discussion of biological citizenship and health resonates with Petryna, *Life Exposed*.

33. Nelson, *Body and Soul*, 10 and passim.

34. Nelson, 22.

35. For example, see Wynter, "Unsettling the Coloniality"; and Wynter and McKittrick, "Unparalleled Catastrophe for Our Species?" In Wynter's framework and following on from Aimé Césaire, the human is "made to the measure of the world." See also Césaire,

Discourse on Colonialism; and Yusoff, *A Billion Black Anthropocenes or None*. In a resonant way, this work suggests that citizens are made to the measure of worlds.

36. While the term "uncommons" surfaces in a broad range of research, de la Cadena has established a particular approach to this concept through pluralistic ontologies. For instance, see de la Cadena and Blaser, *A World of Many Worlds*.

37. This is a recurring thread within pragmatist literature—that pluralism is not a condition of equally available or viable different possibilities, and for this reason it especially points to the power dynamics, modes of participation, and struggles that surface across multiple worlds and world making. For example, see Dewey, *The Quest for Certainty*; and Glaude, *In a Shade of Blue*.

38. The US Environmental Protection Agency has developed an "Air Sensor Toolbox" that includes instructions and protocols on how to collect citizen-sensing data. There are now multiple initiatives to provide more verifiable conditions for citizen data. However, as I discuss in chapter 3, different worlds of data can generate very different ways of working with and mobilizing data. See US Environmental Protection Agency, "Air Sensor Toolbox." For a related discussion of citizen data see Parasie and Dedieu, "What Is the Credibility of Citizen Data Based On?"

39. For instance, *civic* has variously been proposed as an alternative to *citizen*, despite their shared root, since these terms are seen to circulate differently, especially in different national contexts. See Fortun and Fortun, "Scientific Imaginaries and Ethical Plateaus." See also Backstrand, "Civic Science for Sustainability"; and Jasanoff, "Technologies of Humility." There are also now a number of studies on the conjugation of civic and tech, including Boehner and DiSalvo, "Data, Design, and Civics"; and Perng and Maalsen, "Civic Infrastructure."

40. Berlant outlines how worlds might be more binding than expansive: "Optimism is cruel when the object/scene that ignites a sense of possibility actually makes it impossible to attain the expansive transformation for which a person or a people risks striving; and, doubly, it is cruel insofar as the very pleasures of being inside a relation have become sustaining regardless of the content of the relation, such that a person or a world finds itself bound to a situation of profound threat that is, at the same time, profoundly confirming." World binding thus signals the possibility that world making is no guarantee that those worlds will be expansive or enabling. Berlant, *Cruel Optimism*, 2.

41. World Health Organization, "Ambient Air Pollution."

42. Das and Horton note that total deaths from all types of environmental pollution are estimated to be 9 million worldwide, while deaths specifically attributable to air pollution are estimated to be around 6.5 million worldwide. See Das and Horton, "Pollution, Health, and the Planet." They draw their numbers from the Global Disease Burden and the World Health Organization. However, the WHO now estimates that 7 million deaths are due to air pollution per year globally, with 4.2 million premature deaths attributable to ambient air pollution. See World Health Organization, "Air Pollution." Other studies suggest that the actual number of premature deaths due to air pollution could be as high as 8.8 million. See Lelieveld et al., "Cardiovascular Disease Burden."

43. The Royal College of Physicians has provided the estimate of 40,000 deaths attributable to air pollution per year in the UK. See Royal College of Physicians, "Every Breath We Take." The reference to over nine thousand deaths attributable to air pollution per year in London draws on the study from Walton et al., "Understanding the Health Impacts."

44. Berlant, *Cruel Optimism*, 230.

45. Berlant, "Citizenship," 37; see also *Cruel Optimism*, passim. In complement to the earlier-mentioned references on atmospheric citizenship that emphasized air pollution and atmospheres, here I would also add extensive writings on atmospheres as zones of affect. For example, see Choy and Zee, "Condition–Suspension"; Anderson, "Affective Atmospheres"; Adey, "Air/Atmospheres of the Megacity"; McCormack, *Atmospheric Things*; and Stewart, "Atmospheric Attunements." However, as noted elsewhere, I depart from the more Sloterdijkian (as well as Heideggerian) readings of atmospheres that surface in some of these texts.

46. The differential conditions of breathing due to air pollution are a long-standing and recurring topic that environmental justice researchers discuss. For this research I especially draw on Corburn, *Street Science*; Sze, *Noxious New York*; G. Walker, Booker, and Young; "Breathing in the Polyrhythmic City"; Ottinger, "Buckets of Resistance"; and Bullard, "Solid Waste Sites." In addition to the many texts on topics of breathing and atmospheres cited up to this point, a further reference on the making and remaking of bodies includes Górska, *Breathing Matters*. At the same time, Black studies scholars address struggles for breathability in response to structural inequality and racism as they materialize in bodies, relations, atmospheres, and environments. These works further inform this study of citizens of worlds and breathability. See Fanon, *A Dying Colonialism* and *Black Skin, White Masks*; Sharpe, *In the Wake* and "The Weather"; Gumbs, "That Transformative Dark Thing"; and Crawley, *Blackpentecostal Breath*.

47. As Fanon writes, "There is not occupation of territory, on the one hand, and independence of persons on the other. It is the country as a whole, its history, its daily pulsation that are contested, disfigured, in the hope of a final destruction. Under these conditions, the individual's breathing is an observed, an occupied breathing. It is a combat breathing." Fanon, *A Dying Colonialism*, 65.

48. Fanon, 65.

49. In a conversation on John Akomfrah's films, including *Handsworth Songs*, Tina Campt notes in the context of police violence and Covid-19 that "to have the breath be pressed out from you" is to feel the end of the world. See Campt, "John Akomfrah."

50. The notion of atmospheric citizens, as I develop it here, is also less aligned with atmospheric terror or control, which is a prevailing yet somewhat troubling point of reference for its essentialist and undifferentiated discussion that overlooks the actual lived sites of atmospheric violence and unbreathability. This atmospheric-engineering-meets-social-engineering perspective can especially be found within Sloterdijk's militaristic *Terror from the Air*, which is a recurring reference within many atmosphere-focused texts. I depart from this Sloterdijkian influence here to engage instead with research tuned to differential atmospheric inequalities and specific struggles for breathing otherwise. On an expanded discussion of the "otherwise" in relation to breathing, see Crawley, *Blackpentecostal Breath*.

51. See Gumbs, "That Transformative Dark Thing"; and Sharpe, "The Weather."

52. Sharpe, *In the Wake*, 111. For Sharpe, the weather is an atmosphere that is "antiblack" and carries with it the density and memory of slavery, which is pervasive and stifling.

53. Sharpe, 111. For a related discussion that draws on Sharpe, see Simmons, "Settler Atmospherics."

54. Fanon, *Black Skin, White Masks*, 201.

55. Cheah, *What Is a World?*, 199. Citing Fanon, Cheah describes the importance of decolonial struggles that make new subjects and new worlds in which they can flourish, where flourishing and breathing become interchangeable.

56. Somewhat like Cheah (and Fanon), Crawley proposes going beyond a *"merely biological"* approach to breathing, so as to more fully engage with the sociality—and potential openness—of breathing. See Crawley, *Blackpentecostal Breath*, 48.

57. See Dillon and Sze, "Police Power and Particulate Matters," 13.

58. Gumbs, "That Transformative Dark Thing."

59. Gumbs.

60. Gumbs.

61. Gumbs. For a discussion of breathing "after the end of the world," see also Gumbs, *M Archive*.

62. For example, see Kimmerer, *Braiding Sweetgrass.*

63. L. B. Simpson, *As We Have Always Done*, 3.

64. Balibar, *Citizenship*.

65. I especially discuss the co-constitution of bodies and environments in relation to breathing in chapter 6 of *Program Earth*, 162–63. This chapter focuses on sensing air pollution and engages with air and breath as exchanges that undo the hard edges of subjects and milieus. In this analysis, I draw on A. N. Whitehead, *Modes of Thought*, 114.

66. *Citizens of Worlds* is a companion volume to this previous study, since it carries over and extends a Whitehead-inspired approach to designating everything as a subject (including rocks and sensors) that forms as an experiencing entity through connections with environments. As I have previously argued, citizen sensing can mobilize distinct ways of being and becoming a political subject within environmental registers. In this sense, *Program Earth* expands possible designations of citizens across human and more-than-human entities and practices. This text lays the groundwork for developing this analysis into how pluralistic citizens and worlds form.

67. Crawley, *Blackpentecostal Breath*, 1.

68. Law, "What's Wrong with a One-World World?," 130. Law also draws on Anne-marie Mol to discuss how multiple "realities are being done in practices." See Mol, *The Body Multiple*.

69. I develop a theory of practice that draws especially on pragmatism, which I discuss further in the following chapter. By way of comparison, there are a number of theories and investigations of practice that also proliferate within science and technology studies, among other fields. See Gad and Bruun Jensen, "The Promises of Practice"; Rosner, *Critical Fabulations*; and Fortun et al., "Pushback."

70. For an expanded discussion of the many definitions, histories, and uses of "praxis," see Balibar, Cassin, and Laugier, "Praxis."

71. For example, see Pritchard and Gabrys, "From Citizen Sensing to Collective Monitoring."

72. L. T. Smith, *Decolonizing Methodologies*.

73. James, *A Pluralistic Universe*, 68.

74. For a related discussion, see also Mackenzie, *Wirelessness*.

75. See Benjamin, *Race after Technology*.

1. INSTRUMENTAL CITIZENS

1. See Dhanjani, *Abusing the Internet of Things.*

2. A vast range of texts provide resources for social organizing and participation, from the straightforward handbook of Bolton, *How to Resist*, to detailed considerations of how participation is not evenly available, as in Keenaga-Yamahtta Taylor, *How We Get Free*, and how to change the conditions of political engagement through guides that take the form of the syllabus, as in Chrisler, Dhillon, and Simpson, "The Standing Rock Syllabus Project," and the follow-on collection, J. Dhillon and Estes, *Standing with Standing Rock*. For a related critique of discourses of participation, see J. Dhillon's *Prairie Rising.*

3. Ahmed outlines strategies for surviving in a world with which one is at odds, as well as ways to build other worlds, in *Living a Feminist Life.*

4. For just a few examples of these projects, see El Recetario; Makea Tu Vida; and "What You'll Need to Escape New York."

5. See Jencks and Silver, *Adhocism*; Daniek, *Do It Yourself 12 Volt Solar Power.*

6. *The Whole Earth Catalog* is the standard reference here, as discussed in Kirk, *Counterculture Green*; and Turner, *From Counterculture to Cyberculture*. During the early 1970s in Italy, a parallel and radical approach to technology developed with the Global Tools project, documented in Borgonuovo and Franceschini, *Global Tools.*

7. For an example of the serendipity that can emerge through toolkits in the form of Fluxus projects, see Higgins, *Fluxus Experience.*

8. Da Costa and Philip, *Tactical Biopolitics*; Nelson, *Body and Soul.*

9. The Feel Kit is a speculative project from Feel Tank Chicago, described at http://feelkit.feeltankchicago.net/. Ahmed describes the killjoy survival kit in *Living a Feminist Life*, 235–49. The art and research program How to Work Together is an example of a multiplatform collaborative project investigating alternative modes of community organization and collaboration, available at http://howtoworktogether.org/.

10. For instance, see the UK government's Open Policy Making Toolkit.

11. Simondon, *On the Mode of Existence*. Much more could be written about Simondon's specific discussion of instrumentality and the master–slave dynamic within technology, a topic that has also been discussed at length by scholars of race and technology, including Chude-Sokei, *The Sound of Culture* (thanks to Louis Henderson for this reference).

12. Gabrys, "Citizen Sensing: Recasting."

13. Dewey, "The Development of American Pragmatism."

14. James, *Pragmatism and Other Writings*, 27.

15. West, *The American Evasion of Philosophy.*

16. Simondon, *On the Mode of Existence.*

17. This discussion picks up where *Program Earth* left off in thinking about propositions for open technology (which are somewhat different expressions of "openness" from those that call for open hardware, software, and data—since this openness requires attention to the milieus that inform technology as an expanded field of relations).

18. For example, see Noble, *Algorithms of Oppression.*

19. For instance, see Ruha Benjamin's edited collection *Captivating Technology*, which captures the "carceral techniques" that have been implemented in policing, prisons, surveillance, and profiling and yet are also critically engaged with to forge potential sites of retooling and liberation.

20. For examples of these guides, see Tarantola, "How to Erase Yourself from the Internet"; LA Crypto Crew, "How to Become Anonymous Online"; Tactical Tech, "Data Detox Kit"; Zetter, "How to Make Your Own NSA Bulk Surveillance System." This is a short list that could be significantly expanded. For instance, see also Bellingcat's multiple how-to guides, including Ruser, "How to Scrape Interactive Geospatial Data."

21. For an expanded discussion and critique of these reductive and expedient sorts of instrumentalization, see Gabrys, "Programming Environments."

22. Create Lab, "CATTfish"; "Flood Network."

23. Radiation Watch, "Pocket Geiger."

24. Spence, "Earthquake/Vibration Sensor."

25. Seeed Studio, "Grove Smart Plant Care Kit for Arduino."

26. OSBeehives, "BuzzBox"; Veith, "AWS IoT and Beehives."

27. For an extensive example of how to build an air-quality sensor, see rawrdong, "How to Build a Portable, Accurate, Low Cost, Open Source Air Particle Counter."

28. For a commentary on how diverse forms of citizen science can align with different political (or apolitical) objectives, see Kuchinskaya, "Citizen Science."

29. For more on the Dustbox, see https://citizensense.net/kits/dustbox-hardware/. Citizen Sense developed a second version of the AirKit project, available at https://citizen sense.net/projects/airkit/.

30. For a more extensive discussion of electronics and obsolescence, see Gabrys, *Digital Rubbish*.

31. See A. N. Whitehead, *Process and Reality*.

32. See Whyte, "Indigenous Women." Within Indigenous cosmology, distributions of spirituality can inform not just what Whyte calls the "instrumental value" of entities like water but also the "intrinsic value" of these entities because of their connection and agency within cosmologies.

33. The difference, however, is that for Whitehead, cosmologies endure in the realm of abstraction and are drawn into the experiences of actual entities. See Whyte.

34. Stengers makes such a move in her multivolume text *Cosmopolitics*, where, through investigating the history and philosophy of science, she demonstrates how scientific and technical practices make particular worlds hold together, and to what effect. Cosmopolitics, then, describes how these systems of technoscientific relations have political effects, and how they come down to earth. Stengers, *Cosmopolitics I* and *II*.

35. The "tear down" is increasingly becoming a method for unpacking technologies to look at their material composition, infrastructural requirements, and extended operational logic. For example, see yang9741, "How to Tear Down a Digital Caliper."

36. In part, I draw here on an argument made by Dourish and Edwards, who discuss how "pre-packaged expectations of usage patterns" might characterize software components in toolkits, yet toolkits also need to be "designed to accommodate the wide range of potential applications and situations in use." See Dourish and Edwards, "A Tale of Two Toolkits," 34.

37. For example, see Southwest Pennsylvania Environmental Health Project, "Citizen Science Toolkit."

38. Simondon, *On the Mode of Existence*.

39. For an extended discussion on the topic of work-arounds, see L. Houston, Gabrys, and Pritchard, "Breakdown in the Smart City."

40. One very thorough and informative guidebook that does not gloss over the many points of consideration of air-quality monitoring is a citizen sensing guidebook published by the US Environmental Protection Agency. See Williams et al., *Air Sensor Guidebook*.

41. Cvetkovich, *Depression*. See also Pritchard, Gabrys, and Houston, "Re-calibrating DIY."

42. Povinelli, "The Toxic Earth."

43. See Ahmed, *Living a Feminist Life*; Berlant, *Cruel Optimism*; Pritchard, "The Animal Hacker."

44. Berlant, *Cruel Optimism*, 6.

45. See Stengers, *Thinking with Whitehead*.

46. Karvinen and Karvinen, *Getting Started with Sensors*, xi.

47. Karvinen and Karvinen, 2.

48. Karvinen and Karvinen, 9.

49. Karvinen and Karvinen, 4.

50. Ratto and Boler, *DIY Citizenship*, 5.

51. For a critique of the master–servant relationship in technology, see Simondon, *On the Mode of Existence*. For more on a discussion of decolonizing mastery, see Singh, *Unthinking Mastery*.

52. Austin, *How to Do Things with Words*.

53. Latour, "The Berlin Key."

54. Butler, *Bodies That Matter* and *Excitable Speech*.

55. Barad, *Meeting the Universe Halfway*.

56. Isin and Ruppert, *Being Digital Citizens*.

57. For example, see Haraway, *Modest_Witness@Second_Millennium*.

58. See Gabrys, "Sensing Air and Creaturing Data."

59. Wicked Device, "Air Quality Egg."

60. For a discussion of these inscrutable aspects of the Air Quality Egg, see Nold, "Device Studies of Participatory Sensing."

61. For an in-depth ethnography of the communities involved with making and testing the Air Quality Egg, see Zandbergen, "'We Are Sensemakers.'"

62. This process might align with what Suchman, Trigg, and Blomberg have discussed as co-constituted courses of "instructed action" in relation to the prototype. With the prototype, the "configuring" of devices and actions, working practices and sociomaterial relations, is one that relays "across sites of technology development and use." See Suchman, Trigg, and Blomberg, "Working Artefacts," 168.

63. Suchman, *Human–Machine Reconfigurations*, 8–9.

64. Suchman, 22.

65. Wicked Device, "Air Quality Egg."

66. Swart, "Egg Version One End of Life."

67. Latour demonstrates how the how-to, as a form of instruction, can proliferate through scientific infrastructures that travel along with artifacts to ensure that they are suitably encountered. For instance, in relation to a natural history museum collection, he writes, "Even those elements which can withstand the trip, like fossils, rocks or skeletons, may become meaningless once in the basement of the few museums that are being built in the centers, because not enough context is attached to them. Thus, many inventions have to be made to enhance the mobility, stability and combinability of collected items.

Many instructions are to be given to those sent around the world on how to stuff animals, how to dry up plants, how to label all specimens, how to name them, how to pin down butterflies, how to paint drawings of the animals and trees no one can yet bring back or domesticate." Latour, *Science in Action*, 225.

68. Gabrys, "Air Walk."

69. I discuss this project in more detail in the following chapter. For a related discussion, see also Pritchard and Gabrys, "From Citizen Sensing to Collective Monitoring."

70. Citizen Sense, "Citizen Sense Kit."

71. Dewey, *Logic*, 78.

72. A. N. Whitehead, *Process and Reality*, 11.

73. A. N. Whitehead, 11.

74. Benjamin, *People's Science.*

75. Shapin and Schaffer, *Leviathan and the Air-Pump.*

76. Haraway, *Modest_Witness@Second_Millennium*, 23–45.

77. Descartes, *Discourse on Method and Related Writings.*

78. Simondon, *On the Mode of Existence.*

79. Haraway, *Simians, Cyborgs, and Women*, 202–16.

80. For example, see Daston and Galison, *Objectivity*; Pickering, *The Mangle of Practice*; Mody, *Instrumental Community*; and Taub, "Introduction."

81. Latour, "Tarde's Idea of Quantification."

82. As I discuss in *Program Earth*, Simondon describes how information gives form across material, experiential, and epistemic registers through the term "in-form." For example, see Simondon, *Individuation.*

83. As scholars in science and technology studies have noted, many of these experimental engagements are already underway, remaking technologies in practice. For example, see de Laet and Mol, "The Zimbabwe Bush Pump"; and Helmreich, "Reading a Wave Buoy."

84. To this end, works that examine the inequalities that digital technologies operationalize often also propose "otherwise" engagements across subjects, devices, environments, and relations. For example, see Benjamin, *Race after Technology*; and Kukutai and Taylor, *Indigenous Data Sovereignty.*

85. Simondon, *On the Mode of Existence*, xii–xiii.

86. Simondon, 16.

87. A. N. Whitehead, *Process and Reality*, 11.

88. A. N. Whitehead, 11.

89. I discuss the environments needed to sustain facts in a related register in "Sensing Air and Creaturing Data."

90. This is what Whitehead refers to as the "impossibility of tearing a proposition from its systematic context in the actual world." See *Process and Reality*, 11.

91. Reardon et al., "Science and Justice," 13.

92. Reardon et al., 13.

93. Barad, *Meeting the Universe Halfway*, 170.

94. For a discussion of critiques of instrumental reason and instrumental control through the works of Heidegger and Habermas, see Feenberg, *Questioning Technology.* Also extending the Heideggerian consideration of technology and instrumentality, Arendt takes up this topic in relation to her conception of *Homo faber* in *The Human Condition.*

However, in this work I develop a different approach to the instrumental by revisiting and reworking instrumentalism as continuous with experimentalism, as discussed by pragmatist scholars.

95. Barad, *Meeting the Universe Halfway*.

96. As Dewey further writes, "Instrumentalism is an attempt to establish a precise logical theory of concepts, of judgments and inferences in their various forms, by considering primarily how thought functions in the experimental determinations of future consequences." Writing also about the work of James, Dewey suggests that the "reconstructive or mediative function ascribed to reason" becomes a way to develop "a theory of the general forms of conception and reasoning." This suggests that the experimental processes of instrumental concepts are the means by which theories cohere into general forms rather than instrumental approaches proving *a priori* truths. Instrumentalism in this rendering is necessarily experimental and contingent. See Dewey, "The Development of American Pragmatism," 14.

97. Glaude extends pragmatism's discussion of struggle by quoting James, who writes, "The actually possible in this world is vastly narrower than all that is demanded; and there is always the pinch between the ideal and the actual which can only be got through by leaving part of the ideal behind." As Glaude elaborates, this "pinch is a constitutive feature of the world of action," which consists of what James refers to as "'the struggle and the squeeze.'" Glaude, *In a Shade of Blue*, 20–21; James, "The Moral Philosopher," 202–3, 209. Struggles, as I further suggest here, involve tussling across not only ideas and action but also across multiple subjects, modes of political engagement, lived experiences, and worlds in the making.

98. See Dewey, *The Public and Its Problems*.

99. West, *The American Evasion of Philosophy*, 5.

100. James, *Pragmatism and Other Writings*, 27.

101. James, 27.

102. James, 28–30. See also West, *The American Evasion of Philosophy*.

103. This discussion connects to theories of breathing discussed in the Introduction, including combat breathing and breathing as sociality. See Fanon, *A Dying Colonialism*, 65; Gumbs, "That Transformative Dark Thing"; Sharpe, *In the Wake*; and Crawley, *Blackpentecostal Breath*.

104. Ian Hacking notes that instrumentalism came to suggest a certain "antirealism" within the philosophy of science. As West has pointed out, however, pragmatists such as a Dewey, James, and Peirce worked with realist ontologies while evading the fundamental epistemological concerns of philosophies pertaining to truth. In West's analysis, this sidestepping (or "evasion") of epistemologies of truth does not make pragmatism antirealist; rather, it contributes to the evasion of Cartesian framing and concerns with knowing how the real is really real through a preconceived division of subjects and objects. See Hacking, *Representing and Intervening*; and West, *The American Evasion of Philosophy*.

105. For a related discussion on the "work" of political, collective, and democratic life, see Pritchard and Gabrys, "From Citizen Sensing to Collective Monitoring."

106. For example, see Corburn, *Street Science*; and C. M. Dhillon, "Using Citizen Science in Environmental Justice."

107. See Plume Labs, "Clean Air, Together."

108. For analysis of these different ways of parsing the state, the community, and the citizen through or beyond the family, see Berlant, *The Queen of America.*

109. As I noted in the Introduction, a popular reference for discussing air pollution and air control, Sloterdijk's work nevertheless strikes an essentialist and deterministic note in its rendering of the air as a space of terror and control. This study deliberately sidesteps this more fixed reading of air as an "element" "essential" for life, not least because of the rigid political imaginaries that issue forth along with these atmospheric ontologies. See Sloterdijk, *Terror from the Air;* and Müller, "Behind the New German Right."

110. Simondon, *On the Mode of Existence,* 51.

111. Majaca and Parisi, "The Incomputable and Instrumental Possibility."

112. Majaca and Parisi, 1–3.

113. While Dewey opted to use the terms *instrumentalism* and *pragmatism* interchangeably, he also worked with *experimentalism* as a term and concept that attempted to explain the ideas he was developing. In science and technology studies, experiments and experimentality are frequently discussed to describe how these open-ended practices of inquiry and engagement take place. Drawing on Dewey, Ana Delgado and Blanca Callén investigate DIY biology and electronic waste hacking experiments to consider how "hacks" as an "experimental mode of inquiry" open up new approaches to problems. See Delgado and Callén, "Do-It-Yourself Biology." See also Lezaun, Marres, and Tironi, "Experiments in Participation."

114. James, *Pragmatism and Other Writings,* 28 (emphasis in original).

115. Dewey, "The Development of American Pragmatism," 20.

116. As West has pointed out, it is worth noting how experimentalism emerged in pragmatist thought, where the scientific method was seen to be a paragon of "critical intelligence," and experimentalism was very much a product of this practice—where the "social base" for such pragmatism required a more elite professional class to engage in such practices. Nevertheless, West suggests that a possibility for "creative democracy" might still persist in relation to experimentalism. See West, *The American Evasion of Philosophy,* 62, 90, 97, 103.

117. James quoted in Dewey, "The Development of American Pragmatism," 6.

118. James, *Essays in Radical Empiricism.*

119. Haraway, *Modest_Witness@Second_Millennium,* 37.

120. For a related discussion, see DiSalvo et al., "Toward a Public Rhetoric."

121. Peirce is generally credited with having developed the notion of community of inquiry. Peirce developed this idea in relation to the pursuit of logic and science, but pragmatists (especially Dewey) have adapted the concept in relation to democratic modes of inquiry. As Peirce writes, "Unless we make ourselves hermits, we shall necessarily influence each other's opinions; so that the problem becomes how to fix belief, not in the individual merely, but in the community." See Peirce, "The Fixation of Belief."

122. Dewey, *The Public and Its Problems.*

123. Working within a different context, Grant Wythoff describes how "communities of amateur tinkerers" experiment with technologies and gadgets to become the "engine of emerging media." Experimental inquiry in this more hands-on sense becomes part of the process whereby technologies further develop and concretize as media. See Wythoff, *The Perversity of Things,* 37.

124. Spivak, *Imperatives to Re-imagine the Planet.*

125. Simpson further writes, "engagement changes us because it constructs a different world within which we live." See L. B. Simpson, *As We Have Always Done*, 19–20.

126. These questions also introduce what science and technology studies scholars have referred to as the "politics of how." See Dányi, "The Politics of 'How'"; and Law and Joks, "Indigeneity, Science, and Difference."

127. These modes of collaboration could be described as "contingent collaborations" that work across different engagements, here with environmental conflict and social justice. See Tuck et al., "Geotheorizing Black/Land."

128. Irani, "Hackathons," 807.

129. L. B. Simpson, *As We Have Always Done*, 23.

130. Simpson, 20.

131. While James works with the notion of the "pluralistic universe," *pluriverse* is a term that other writers such as Latour have used in relation to James's work. Writers such as Walter Mignolo take up the pluriverse as a concept but do not cite James as part of its development, instead developing *pluriverse* as a term associated with postcolonial and decolonial theory. Marisol de la Cadena develops yet another reading of the pluriverse; initially through the conservative political writings of Carl Schmitt. This study recognizes these multiple formations of the pluriverse (and somehow it is fitting that this term has a plurality of uses and affiliations) but especially emphasizes James's discussion of the pluralistic universe. See James, *A Pluralistic Universe*; Latour, *Facing Gaia*, 36; Mignolo, "On Pluriversality"; and de la Cadena, "Indigenous Cosmopolitics in the Andes."

132. For a discussion of collective causation in relation to environmental protest, see Zitouni, "Planetary Destruction."

133. Haraway, *Modest_Witness@Second_Millennium*, 23–45.

134. Barad, *Meeting the Universe Halfway*, 170.

135. Dewey, "The Naturalization of Intelligence."

136. West, *The American Evasion of Philosophy*, 82.

137. For a discussion of how to decolonize methodologies, along with an extensive set of case studies and examples of community research, see L. T. Smith, *Decolonizing Methodologies*.

138. For a discussion of this question, see Gane and Haraway, "When We Have Never Been Human."

139. Berlant, *Cruel Optimism*, 3.

140. Stengers, "Including Nonhumans in Political Theory."

141. For instance, see Pollock and Subramaniam, "Resisting Power, Retooling Justice"; and Ahmed, *Living a Feminist Life*.

142. For a discussion of the limits of categories in relation to citizen science, see Irwin, "Citizen Science and Scientific Citizenship."

143. Center for Urban Pedagogy, "Making Policy Public."

144. Hickey, *A Guidebook of Alternative Nows*.

145. Blas, *Gay Bombs*.

146. Allahyari and Rourke, *The 3D Additivist Cookbook*.

147. Detroit Community Technology Project. See also Institute of Technology in the Public Interest.

148. There is insufficient space here to cover the proliferation of different forms of toolkits, handbooks, and guides. A notable example at the intersection of technology and

democracy is Soon and Cox, *Aesthetic Programming*. At the intersection of environmental justice and air-pollution monitoring, the West Oakland Environmental Indicators Project provides many excellent resources for monitoring and improving environments. See https://woeip.org.

149. Nelson, *Body and Soul*.

150. Dewey, "The Development of American Pragmatism," 12.

151. Berlant, *Cruel Optimism*, 21.

2. SPECULATIVE CITIZENS

1. Moore, "Air Impacts."

2. International Agency for Research on Cancer, World Health Organization, "IARC."

3. Climate and Clean Air Coalition, "Satellite Data"; and Zhang et al., "Quantifying Methane Emissions."

4. This particular video can be found at https://www.youtube.com/watch?v=OQmoK1DJIyE.

5. This FLIR technique has become much more pervasive across environmental NGOs and journalism. For instance, see Kessel and Tabuchi, "It's a Vast, Invisible Climate Menace"; and Environmental Defense Fund, "With This Technology." The latter refers to methane leaks as "oil spills in the sky."

6. Frank Finan's YouTube channel is https://www.youtube.com/channel/UC7Eph33czawYR2ZKZrexSoQ. See also Vera Scroggins's YouTube channel, https://www.youtube.com/channel/UCfbfGPFJn5t3HvJcRNTvJxQ.

7. World Health Organization, "Ambient (Outdoor) Air Quality and Health."

8. For instance, see Bullard, "Solid Waste Sites"; Buzzelli et al., "Spatiotemporal Perspectives"; Corburn, *Street Science*; Dillon and Sze, "Police Power and Particulate Matters"; Mitchell and Dorling, "An Environmental Justice Analysis"; Morello-Frosch, Pastor, and Sadd, "Environmental Justice and Southern California's 'Riskscape'"; and Pearce, Kingham, and Zawar-Reza. "Every Breath You Take?" For an example of an especially pluralistic approach to spatial distribution and environmental injustice, see G. Walker, "Beyond Distribution and Proximity."

9. Sharpe, *In the Wake*, 109. "Breathtaking" is a term also used by Alison Kenner (although not in reference to Sharpe) in a study of asthma. See Kenner, *Breathtaking*.

10. More information on fracking exemptions from environmental safeguards can be found at National Resources Defense Council, "NRDC Policy Basics." A discussion of federal- and state-level oil and gas regulations is available at Phillips, "Burning Question." The US Energy Policy Act of 2005 can be found at https://www.gpo.gov/fdsys/pkg/PLAW-109publ58/html/PLAW-109publ58.htm. Energy policies continue to vacillate in relation to different political agendas, with the forty-fifth US president significantly rolling back regulations on oil and gas. For example, see Geiling, "Trump Administration Officially Scraps Obama-Era Rules."

11. Ingraffea et al., "Assessment and Risk Analysis"; Osborn et al., "Methane Contamination of Drinking Water." For an example of a well organized citizen-based water-monitoring initiative in Pennsylvania, see the Alliance for Aquatic Resource Monitoring, along with their manual, "Shale Gas Extraction."

12. There is an extensive and varied array of academic research on fracking. For example, see Green, "Fracking the Karoo"; Kama, "Resource-Making Controversies"; Kinchy,

Jalbert, and Lyons, "What Is Volunteer Water Monitoring Good For?"; Lave and Lutz, "Hydraulic Fracturing"; Neville et al., "Debating Unconventional Energy"; Willow et al., "Contested Landscape"; and Wylie, *Fractivism*. While this chapter does not have space to survey this broad-ranging work, it does align with studies specifically oriented to how energy and extraction spark or constrain political possibilities. In this sense, this discussion resonates with broader developments and investigations into energy as a social, political, material, and cultural organizing force. For example, see Szeman and Boyer, *Energy Humanities*; and Barry, *Material Politics*.

13. For an example of other modes of science (fiction) within Indigenous practices of observation and narration, see Whyte, "Indigenous Science (Fiction)."

14. "Care about your air" is a strapline on the box of the Air Quality Egg product, a relatively well known DIY air-quality monitor available for purchase (discussed in more detail in chapter 1). "Care for Your Air" is also a motto for the EPA in its indoor air-quality initiative. See US Environmental Protection Agency, "Care for Your Air." It is also the name for an air-pollution initiative in India, Care for Air.

15. Berlant, *Cruel Optimism*.

16. Puig de la Bellacasa, "Matters of Care in Technoscience," 85. See also Mol, *The Logic of Care*; and Mol, Moser, and Pols, "Care."

17. I discuss this point further in *Program Earth*. This extended discussion also draws on Stengers, *Thinking with Whitehead*; and A. N. Whitehead, *Process and Reality*.

18. Stengers, *Thinking with Whitehead*, 267; and A. N. Whitehead, *Process and Reality*, 11.

19. Shaviro, *Without Criteria* and *The Universe of Things*.

20. See Gill, Singleton, and Waterton, "The Politics of Policy Practices."

21. In this way, Citizen Sense worked with communities to understand and develop ways of documenting pollution and proposing different modes of action together, based on preexisting community practices. This approach builds on attempts to decolonize research methodologies. See L. T. Smith, *Decolonizing Methodologies*.

22. State Impact, "The Marcellus Shale Explained"; and Soeder and Kappel, "Water Resources and Natural Gas Production."

23. Fractracker Alliance, "Pennsylvania Shale Viewer."

24. Griswold, "The Fracturing of Pennsylvania." For an extended investigation, see Griswold, *Amity and Prosperity*.

25. Griswold, "The Fracturing of Pennsylvania."

26. Llewellyn et al., "Evaluating a Groundwater Supply Contamination."

27. Some estimates indicate that up to 750 chemicals are used in the fracking process, many of which are also endocrine disrupters. For instance, see Kassotis et al., "Estrogen and Androgen Receptor Activities." However, not all chemicals are used at the same time or place. Other sources suggest that "50 known chemicals" "may be added to the water that is used for hydraulic fracturing." See http://exploreshale.org. These lists of chemicals are obtained from industry sources, which might not disclose (since they are not required to) all chemicals, particularly proprietary chemicals, used in the fracking process. For instance, see Department of Environmental Protection Bureau of Oil and Gas Management. "Chemicals Used by Hydraulic Fracturing Companies."

28. Howarth, Santoro, and Ingraffea, "Methane and the Greenhouse-Gas Footprint."

29. Gabrys and Yusoff, "Arts, Sciences, and Climate Change"; Latour, "Atmosphère, Atmosphère"; Briggle, *A Field Philosopher's Guide*.

30. There is extensive research on this problem. See especially Griffith Spears, *Baptized in PCBs*.

31. Chen, *Animacies*; Schrader, "Responding to *Pfiesteria piscicida*."

32. Felt and Fochler, "Bottom-up Meanings."

33. Cohen, "Challenges and Benefits."

34. Edwards, "Pigeon Air Patrol."

35. Gabrys, Pritchard, and Barratt, "Just Good Enough Data"; US Environmental Protection Agency, "Draft Roadmap."

36. Stengers, *Thinking with Whitehead*, 518; Gabrys, *Program Earth*.

37. Griswold, "The Fracturing of Pennsylvania."

38. The "List of the Harmed" is an ongoing record last updated August 22, 2018, to which anyone can contribute by emailing the list moderator. It is available at http://pennsylvaniaallianceforcleanwaterandair.wordpress.com/the-list.

39. Olsen, "Natural Gas and Polluted Air." This journalist's reference to incomplete science in some ways resonates with "undone science" and knowledge struggles as discussed in Frickel et al., "Undone Science."

40. Earthworks, "Gas Patch Roulette."

41. Macey et al., "Air Concentrations of Volatile Compounds."

42. See Fractracker Alliance; and Marcellus Gas.

43. Roter, "Breathe Easy Susquehanna County."

44. For an overview of these different techniques, see Steinzor, "Community Air Monitoring."

45. Murphy, *Sick Building Syndrome*. As Corburn importantly points out, however, there are many possible ways of engaging with citizen data beyond the framings of epidemiology and risk. Such an expanded approach would work to ensure that citizen data could operate in more and other registers than those in direct comparison to expert methods. See Corburn, *Street Science*, 5–10.

46. Murphy, *Sick Building Syndrome*, 81–110. See also Irwin, "Citizen Science."

47. As Stengers notes in relation to Whitehead, practices of perceptibility and imperceptibility can indicate how "the future hesitates in the present." Stengers, *Thinking with Whitehead*, 191.

48. There are six criteria pollutants that the EPA monitors because they are especially hazardous to human health, environments, and organisms: carbon monoxide, lead, ground-level ozone, particulate matter, nitrogen dioxide, and sulfur dioxide. More information is available at https://www.epa.gov/sites/default/files/2015-10/documents/ace3_criteria_air_pollutants.pdf.

49. For more information on the Citizen Sense Toolkit, see https://citizensense.net/kits/citizensense-kit. Frackbox assembly instructions are also available at https://citizensense.net/kits/frackbox-hardware.

50. See Create Lab, "Speck." The Heinz Foundation supported the distribution of Specks in libraries. See Spice and Thinnes, "CMU, Airviz."

51. For related approaches to the problem of observation and action, see Dewey, "The Development of American Pragmatism"; and Haraway, *Modest_Witness@Second_Millennium*.

52. Citizen Sense, "Citizen Sense Monitoring Events."

53. For an extended discussion of these aspects of the participatory research process, see Pritchard and Gabrys, "From Citizen Sensing to Collective Monitoring."

54. On these possibilities for other sciences and knowledge practices, see Jasanoff, "Technologies of Humility"; Benjamin, *People's Science*; McKittrick, *Dear Science and Other Stories*; and Rusert, *Fugitive Science*.

55. For an expanded discussion of the Citizen Sense walking methodologies developed during the research (along with other approaches to walking-as-research that inspired this approach), see Gabrys, "Air Walk."

56. See Colaneri, "Gas Drilling Draws Citizen Scientists."

57. openair was developed by David Carslaw. More information is available at https://davidcarslaw.github.io/openair/.

58. To read the five data stories see Citizen Sense, "Pennsylvania Data Stories." For project films see Citizen Sense, "Pollution Sensing Videos."

59. Cox, "Citizens' Digital Monitoring Project."

60. Agency for Toxic Substances and Disease Registry, "Health Consultation."

61. Pennsylvania Department of Environmental Protection, "DEP Expands Particulate Matter Air Monitoring Network."

62. Hurdle, "PA Expands Particulate Monitoring."

63. Breathe Easy Susquehanna Co., "Citizen Science." After this study, citizens and community groups also purchased several Specks, which they made available in the public library for general use.

64. The full list of pollutants monitored at this site, including carbonyls and VOCs, can be found at: https://www.facebook.com/BreatheEasySusq/posts/citizen-science-spurred-real-air-quality-monitoring-in-the-pa-shalegas-fields-re/895271397320857/.

65. Davenport, "Trump Eliminates Major Methane Rule."

66. Hurdle, "PA Expands Particulate Monitoring."

67. See West, *The American Evasion of Philosophy*, 147–48. West writes about the development of W. E. B. Du Bois's pragmatism and his sense of the limitations of individual—in comparison to collective—pursuits within a creative democracy.

68. Writing about the uneven qualities of the empirical, Benjamin notes that "demanding empirical evidence of systematic wrongdoing can have a kind of perverse quality—as if subjugated people must petition again and again for admission into the category of 'human,' for which empathy is rationed and applications are routinely denied." Benjamin, "Racial Fictions, Biological Facts," 2.

69. Benjamin, 2.

70. See Lane et al., "Doing Flood Risk Science Differently;" Waterton and Tsouvalis, "'An Experiment with Intensities.'"

71. Dewey, "The Development of American Pragmatism."

72. West, *The American Evasion of Philosophy*, 91.

73. See Pritchard and Gabrys, "From Citizen Sensing to Collective Monitoring"; Combes, *Gilbert Simondon*, 34–35.

74. Or as Stengers writes, this would involve the "invention of the field in which the problem finds its solution." Stengers, *Thinking with Whitehead*, 17.

75. Stengers, 17.

76. Gabrys, "A Cosmopolitics of Energy" and "For the World." See also Gill, "Caring for Clean Streets."

77. Stengers, *Thinking with Whitehead*, 147; A. N. Whitehead, *Science and the Modern World*, 86.

3. DATA CITIZENS

1. The literature on the health effects from air pollution is vast. One current study estimates that as many as 8.8 million deaths worldwide are due to air pollution each year. See Lelieveld et al., "Cardiovascular Disease Burden"; and Gabrys, "Planetary Health in Practice."

2. In 2005 the WHO established guidelines for $PM_{2.5}$, including 25 µg/m³ for twenty-four-hour mean and 10 µg/m³ for the annual mean. The guidelines have since been updated in 2021 to 15 µg/m³ for twenty-four-hour mean and 5 µg/m³ annual mean. However, as health research on air pollution notes, there is no safe level of exposure to $PM_{2.5}$. See World Health Organization, *WHO Air Quality Guidelines;* World Health Organization, "Ambient (Outdoor) Air Pollution"; Dockery et al., "An Association between Air Pollution and Mortality"; Grigg, "Where Do Inhaled Fossil Fuel–Derived Particles Go?"; and Holgate, "Every Breath We Take."

3. Analyses of air-pollution events as they inform citizens' engagement with environments, especially through digital technologies, are now increasingly common. For instance, in the Chinese context see Kay, Zhao, and Sui, "Can Social Media Clear the Air?"; Li and Tilt, "Public Engagements with Smog"; and Aunan, Hansen, and Wang, "Introduction."

4. As reported in "India: Health Emergency Declared as Toxic Air Shrouds New Delhi." Based on this story, it is unclear which pollutants measured "999" on the Air Quality Index (AQI). The AQI is available at https://aqicn.org/city/delhi.

5. This work is situated within long-standing environmental justice research that studies these distributions of inequality and pollution. See Bullard, *Dumping in Dixie;* Sze, *Noxious New York*; and Corburn, *Street Science.*

6. For a related discussion that addresses this unequal distribution not just within cities but also across countries, see Hecht, "Air in a Time of Oil."

7. As UN Special Rapporteur David Boyd writes of the 2019 report on the right to breathe clean air, "The Special Rapporteur focuses on the right to breathe clean air as one of its components and describes the negative impact of air pollution on the enjoyment of many human rights, in particular the right to life and the right to health, in particular by vulnerable groups. He highlights the different state obligations in relation to the right to breathe clean air, which are both procedural and substantive, as well as the specific obligation to protect people and groups in vulnerable situations." See Boyd, "The Right to Breathe Clean Air." This report sits within a broader framework of more than one hundred countries agreeing to a right to a healthy environment. See United Nations Human Rights Council, "Human Rights and the Environment."

8. Guidelines that establish measures for what counts as clean air include the European Commission Directive 2008/50/EC; and World Health Organization, *WHO Air Quality Guidelines.*

9. For an example of one such challenge, see ClientEarth Communications, "Client Earth Launches."

10. Isin and Ruppert, *Being Digital Citizens.*

11. See Haraway, *Modest_Witness@Second_Millennium*; and Gabrys, Pritchard, and Barratt. "Just Good Enough Data."

12. Alan Irwin discusses how those affected by environmental matters should be involved in decision-making processes as a way to build trust and address ethical concerns. See Irwin, "Citizen Science and Scientific Citizenship."

13. John Law has referred to how people are formatted and enacted through surveys in "What's Wrong with a One-World World?" I have similarly discussed the programming of citizens and environments via sensor technologies in *Program Earth*.

14. Gregory and Bowker. "The Data Citizen," 220.

15. Benjamin, *Race after Technology*.

16. For a more extensive discussion of the pitfalls of the Citizen app, see Ashworth, "Inside Citizen." As the article notes, the Citizen app was originally launched in 2016 under the name Vigilante. As the podcast included as part of Ashworth's article notes, the app currently covers eighteen cities in the United States, but the makers hope to monetize and expand to 1.5 billion users worldwide.

17. The literature on wearables is vast, and I do not have space to discuss this here. However, several researchers address the formations of citizenship and political subjects through wearable sensors. See Lupton, *The Quantified Self*; and Boyle, "Pervasive Citizenship."

18. Literature on these topics is equally vast, and there is no space to engage with the many studies in this dynamic field. A representative range of research on data practices and politics that informs *Citizens of Worlds* includes Milan and Treré, "Big Data from the South(s)"; Couldry and Powell, "Big Data from the Bottom Up"; Loukissas, *All Data Are Local*; Kukutai and Taylor, *Indigenous Data Sovereignty*; and Meng and DiSalvo, "Grassroots Resource Mobilization."

19. For instance, see initiatives such as Data for Black Lives, also discussed elsewhere in this chapter (https://d4bl.org); as well as Currie et al., "The Conundrum"; Gutiérrez, *Data Activism and Social Change*; Bruno, Didier, and Vitale, "Statactivism"; and Renzi and Langlois, "Data Activism."

20. There has been a proliferation of studies that examine power and justice in relation to data. Another partial list of work in this area includes Dencik, Hintz, and Cable, "Towards Data Justice?"; L. Taylor, "What Is Data Justice?"; and D. Walker et al., "Practicing Environmental Data Justice." Working in a different but related register are projects including the Our Data Bodies Project; Onuoha and Mother Cyborg, "People's Guide to AI"; Mertia, *Lives of Data*; and Cifor et al., "Feminist Data Manifesto-No."

21. As Justin Pidot writes, "The new law is of breathtaking scope. It makes it a crime to 'collect resource data' from any 'open land,' meaning any land outside of a city or town, whether it's federal, state, or privately owned. The statute defines the word collect as any method to 'preserve information in any form,' including taking a 'photograph' so long as the person gathering that information intends to submit it to a federal or state agency. In other words, if you discover an environmental disaster in Wyoming, even one that poses an imminent threat to public health, you're obliged, according to this law, to keep it to yourself." Pidot, "Forbidden Data"; see also Kravets, "Law Making It Illegal to Collect Data."

22. US Court of Appeals, Tenth Circuit, "People for the Ethical Treatment of Animals."

23. Rights to participate can be variously recognized, with two notable examples: the 1998 Aarhus Convention (or the UNECE Convention on Access to Information, Public Participation in Decision-making and Access to Justice in Environmental Matters) and the Protocol on Pollutant Release and Transfer Registers. These measures seek to protect the right to environment through the right to participate. See United Nations Economic Commission for Europe, "Public Participation."

24. See Berlant, *Cruel Optimism*, passim.

25. In this sense, this approach departs from the focus on rights as speech acts as discussed by Isin and Ruppert in *Digital Citizens*.

26. These differently constituted and emerging rights feature in Spivak's discussion of migration in *Imperatives to Re-imagine the Planet*.

27. Étienne Balibar configures citizenship along these lines in *Citizenship*.

28. I developed this concept by building on Foucault's notion of environmentality. See Gabrys, "Programming Environments."

29. Balibar, *Citizenship*, 18. Such continued invention of democracy resonates differently with Dewey's pragmatist articulation of political engagement, as well as Black pragmatist and Indigenous discussions of democratic politics as unfolding through struggle, praxis, mutual exchange, and reciprocity. See Dewey, The *Public and Its Problems*; Glaude, *In a Shade of Blue*; and Denzin, Lincoln, and Smith, *Handbook of Critical and Indigenous Methodologies*.

30. This approach resonates with pragmatism's general orientation toward prospective conditions. While Dewey engaged especially in the prospective conditions of democratic participation, James established how forms of knowing and things that are in the making characterize pragmatist methods. He writes: "What really *exists* is not things made but things in the making." James, *A Pluralistic Universe*, 263. Elsewhere, James contrasts rationalism and pragmatism to suggest that rather than being "ready-made," reality "is still in the making, and awaits part of its complexion from the future." David Lapoujade suggests that this approach characterizes pragmatism as a method (something that James also discussed). See Lapoujade, *William James*.

31. This approach to conditions in the making continues within pragmatist work from James to Dewey, who writes of the self and "worlds in the making" through different modes of conduct, deliberation, and conflict. See Dewey, *Human Nature and Conduct*, 150. Also quoted in Glaude, *In a Shade of Blue*, 30.

32. Rights, citizenship, and environmental participation could, by extension, materialize less as fully formed conditions and more as sought-after relations. In a different but resonant register, Henri Lefebvre discusses the right to make the city as a collective work, as the "right to the *ouevre*," which spans the right to public space, difference, housing, political engagement, social life, and even information technology. Such an articulation of the right to the city suggests possible ways to reinvent rights as open-ended and in the making. The pursuit and exercise of the right to the city could occur through attempts not only to make a claim to the city but also to actively shape it as a more breathable world. In this sense, rights could become another sort of instrument that materializes through practices of open-air instrumentalism. See Lefebvre, *Writings on Cities*, 145, 157, 174.

33. Berlant, "Citizenship."

34. Berlant, *Cruel Optimism*, 4. This focus on failure and impasse resonates in a different way with Glaude's conception of tragedy and the blues, which he suggests (drawing

on James) should more fully infuse and be acknowledged as a necessary condition of pragmatism, as unfolding through "the struggle and the squeeze of the world of action." See Glaude, *In a Shade of Blue*, 22.

35. Berlant, *Cruel Optimism*, 227.

36. See Seitz, interview with Berlant. This observation is also included in the introductory epigraph to *Citizens of Worlds*.

37. Writing in a similar way in relation to Black Americans, Alondra Nelson refers to the "citizenship contradiction" that occurs in the "gap between civil rights and social benefits." Nelson, *Body and Soul*, 10.

38. For example, see Kimura and Kinchy, "Citizen Science in Post-Fukushima Japan."

39. Data for Black Lives.

40. Cocco and Smith, "Race and America." This *Financial Times* article includes an interview with Milner, as well as reference to W. E. B. Du Bois's data visualizations, collected in Battle-Baptiste and Rusert, *W. E. B. Du Bois's Data Portraits*.

41. As the Data for Black Lives website notes, "We are a movement of scientists and activists. Data as protest. Data as accountability. Data as collective action."

42. Mayor of London, London Data Store, "2011 Census Ethnic Group Fact Sheets."

43. South East London Combined Heat and Power, "History." For a discussion of community resistance to the incinerator, see Parau and Wittmeier Bains, "Europeanisation as Empowerment."

44. Laville, "UK Waste Incinerators."

45. Evelyn, *Fumifugium*.

46. Steele, *Deptford Creek Surviving Regeneration*.

47. Steele, 2.

48. Steele, v.

49. "Deptford Is Changing" is a visual–social research project and website by Anita Strasser that documents these more current changes. See especially Strasser, "Tidemill Garden."

50. Don't Dump on Deptford's Heart, "Help Us Combat Thames Tunnel Pollution." See also Thames Tideway Tunnel, https://www.tideway.london/the-tunnel/.

51. The results from the diffusion-tube monitoring can be reviewed at the Don't Dump on Deptford's Heart project's Google Map.

52. An extensive background literature informs this discussion on infrastructure, especially in relation to how material and political relations and possibilities of action are distributed through infrastructures. See Berlant, "The Commons"; Bruun Jensen and Morita, "Introduction"; Larkin, "The Politics and Poetics of Infrastructure"; Hawkins, "Governing Litter"; and Maguire and Winthereik, "Digitalizing the State."

53. For a related discussion of this approach as researcher–participants, see Pritchard and Gabrys, "From Citizen Sensing to Collective Monitoring."

54. Loxham, Davies, and Holgate, "Health Effects."

55. Shinyei, Particle Sensor Unit PPD42NJ.

56. For the site data see Department for Environment, Food, and Rural Affairs, "Site Information for London Marylebone Road (UKA00315)."

57. For a more extensive discussion of our approach to calibrating the Dustbox, see Pritchard, Gabrys, and Houston, "Re-calibrating DIY."

58. A video of the workshop and walk is available at Citizen Sense, "Urban Sensing Using the Dustbox," and is also available at https://manifold.umn.edu/projects/citizens -of-worlds/resource-collection/citizens-of-worlds-videos/resource/sensing-dustbox.

59. Before the Citizen Sense monitoring commenced, a monitor sited on Mercury Way near the incinerator and waste-transfer yard was taken offline by Lewisham Council, with the intention of moving this site to the UK Environment Agency. However, Mercury Way does not appear on the Department for Environment, Food, and Rural Affairs' "Interactive Monitoring Networks Map." After the Citizen Sense monitoring had been completed, Lewisham Council brought two new urban-background air-quality monitoring stations online in Deptford, one near the Thames Tunnel Super Sewer construction site and the other at Honor Oak Park (an air-pollution research supersite). See London Air, "Lewisham."

60. For a more extensive discussion of these encounters with troubleshooting, see L. Houston, Gabrys, and Pritchard, "Breakdown in the Smart City."

61. It should be noted that there was no vandalism or theft of the Dustboxes, which remained in place without incident when installed outdoors for monitoring.

62. See also Gabrys, "Planetary Health in Practice."

63. As indicated in the introduction to this chapter, this discussion builds on concepts of figuring developed in Haraway's *Modest_Witness@Second_Millennium* and in Gabrys, Pritchard, and Barratt's "Just Good Enough Data." Figurations are ways of worlding, and they can take the form of stories and numbers. Indeed, numbering is a way of narrating. Sociological studies of quantification are extensive and it would not be possible to cover the richness of this area of analysis here. However, several aligned texts that inform this study include Verran, "The Changing Lives"; Asdal, "Enacting Things through Numbers"; and Lippert and Verran, "After Numbers?" Enumeration is a process and practice for making objects, infrastructures, and governance practices, but as these texts show, it is rarely if ever a simple process of counting or accounting and instead involves social-calculative relations and practices.

64. In an examination of sensing air quality in *Program Earth* I suggested that it could be possible to engage with data less as free-floating facts or as the monolithic products of expertise and more as *creatures* that are constituted with and through environments of relevance. Air pollution, in this way, is constituted through numbering practices that configure and creature air pollution as a specific object of relevance. Different modalities of data in turn can generate different figurations and creatures of air pollution. In developing this analysis, I draw on Alfred North Whitehead's discussion, in *Process and Reality*, of creatures as the actual entities and occasions that coalesce through processes and relations.

65. Different approaches to narratives, storying, and fictions surface through these practices, that could follow much different trajectories of "science," observation, and experience as indicated in work ranging from McKittrick's *Dear Science and Other Stories* to Nadim's "Blind Regards." Storytelling can also be a way to hold a plurality of experiences together without resolving them, as noted in D. Houston, "Environmental Justice Storytelling"; and Spencer, Dányi, and Hayashi, "Asymmetries and Climate Futures."

66. The *Deptford Data Stories* are available at https://citizensense.net/data-stories -deptford.

67. A resonant discussion of stories can be found in Petryna's *Life Exposed*, xxvi.

68. Morgan, "'It's Time to Act Now.'"

69. Foxcroft, "Business of the House."

70. Foxcroft.

71. The latest state of plans for this site can be found at Peabody, "Frankham Street Development." For an account of the site transformation and community protest at the development, see Worthington, "Deptford's Tidemill Campaign."

72. For an example of some of these initiatives, see Save Reginald! Save Tidemill!, "Help Us Save Reginald House and Tidemill Wildlife Garden."

73. Corporate Watch, "Tidemill."

74. Vickers, "The Battle for Deptford."

75. Cuffe, "Lewisham Paying Back Debt."

76. Save Reginald! Save Tidemill!, "Destruction of Deptford's Much Loved Community Garden."

77. Noor, "Housing Approved Despite Pollution Warning."

78. Crosswhatfields?, "No. 1 Creekside."

79. After extensive protest over development plans at the Old Tidemill site, Peabody incorporated a community-consultation process as part of the development. They write: "Our aim is for everyone to participate in the design of the open and public spaces that will be delivered within the new scheme. . . . We have consulted with the community on the designs for the green space and have been working with a small group of local people to shape the proposals." No indication is given as to how this small group was selected, whom it involves, or to what extent it is representative of local interests. See Peabody, "Frankham Street Development."

80. Virginia Eubanks suggests that these analyses of power must remain a critical component of participatory research. See Eubanks, "Double-Bound."

81. Rankine, *Citizen*. Balibar also notes how civility and political participation can be at odds when attempting to undertake democratic engagement and challenge existing power structures. See Balibar, *Citizenship*, 53 and passim.

82. Berlant, *Cruel Optimism*, 28.

83. Nelson, *Body and Soul*.

84. Thanks to Helen Pritchard for this discussion about countering the perennial promise of making and doing, which can often foreclose the necessity of unmaking and undoing. These practices are likely to occur in conditions of conflict and struggle, since they work against rather than reinforce established technoscientific practices. See also McGlotten, "Black Data."

85. Glaude discusses this point at length in relation to Dewey and pragmatism in *In a Shade of Blue*.

86. The Deptford Park area is located within Evelyn Ward, one of the most impoverished wards in the UK. See End Child Poverty.

87. Deptford Folk.

88. Royal College of Physicians, "Every Breath We Take."

89. Deptford Parks.

90. As Lewisham Council writes, "We have expanded our network of air quality monitoring. A new site has been set up in Deptford which increases the continuous monitoring sites to four. A new state of the art supersite has recently been set up at Honor Oak Park sportsground. This includes important research being carried out by Kings College

London." The council's efforts especially focused on engaging with atmospheric scientists who had a preestablished relationship with the council. See Lewisham Council, "What We Are Doing."

91. Hancock, "Khan Calls Lewisham Emissions 'Health Crisis.'"

92. Following on from James and Dewey, especially as read by Glaude, contingency and action here characterize open-air instrumentalisms through the maxim "Act, but at your peril." This statement from Dewey indicates how practical activity guided toward change can generate uncertain effects in the making and remaking of worlds. See Dewey, *The Quest for Certainty*, 6; and Glaude, *In a Shade of Blue*, 22.

93. The Ella Roberta Family Foundation.

94. The Ella Roberta Family Foundation, "About the Foundation."

95. PA Media, "Inquest"; and Laville, "Air Pollution a Cause." See also Blackstone Chambers, "Inquest into the Death of Ella Adoo-Kissi-Debrah"; and London Inner South Coroner's Court, "Inquest Touching the Death of Ella Roberta Adoo Kissi-Debrah."

96. The Ella Roberta Family Foundation's "Every Breath Matters" is a film that narrates the effects of air pollution on Ella and on 93 percent of children around the world. It ends with the appeal to rights, "Clean Air Is a Human Right," as well as the hashtag #EveryBreathMatters, a campaign to demonstrate how every breath has an accumulative and potentially lethal effect.

97. Tobin, "Extinction Rebellion Lewisham."

98. Carrington, "Covid-19 Impact on Ethnic Minorities."

99. As noted in the Introduction, Fanon wrote, "It is not because the Indo-Chinese has discovered a culture of his own that he is in revolt. It is because 'quite simply' it was, in more than one way, becoming impossible for him to breathe." Fanon, *Black Skin, White Masks*, 201. This quote, and paraphrases of it, have become a common refrain in Black Lives Matter, as well as environmental and social-justice movements and actions.

100. Mbembe, "The Universal Right to Breathe."

101. Hannah Arendt famously raised this line of critique in her discussion of the "right to have rights," which would require a (universal) governing body to ensure the realization of rights. See Arendt, *The Origins of Totalitarianism*; and DeGooyer et al., *The Right to Have Rights*. A different but resonant critique of universality vis-à-vis pragmatism can be found in Glaude's *In a Shade of Blue*, which calls attention to the specificity of justice and ethics. These works, however, do not create a simple binary between the universal and the situated, but rather emphasize the struggles to realize principles or rights that might be articulated in a more universal register. In so doing, they raise the challenge of how to realize justice through, and not despite, these struggles.

102. See also Gabrys, Pritchard, and Barratt, "Just Good Enough Data."

103. Benjamin, "Racial Fictions, Biological Facts," 2. As discussed in chapter 1 in a related register, Cornel West raises a critique of Deweyan pragmatism and its possible reliance on the scientific method as the basis for democracy. See West, *The American Evasion of Philosophy*. Writing along with these texts, I suggest here that pluralistic data practices can make multiple worlds and activate citizenships, where the right to data also forms the right to experience. In other words, many other data ontologies could surface through different formations of evidence and experience.

104. See also Stengers, *Thinking with Whitehead*, passim.

4. MULTIPLE CITIZENS

1. For example, see Howe, "Sensing Asymmetries"; and Gramaglia and Mélard, "Looking for the Cosmopolitical Fish."

2. I have previously written about this topic, which I continue to develop through this discussion of air-quality gardens. See Gabrys, "Becoming Urban."

3. For a related discussion on how to move beyond human-based understandings of sensing and embodiment, and what these transformations of sense can generate (especially in relation to aquifers and satellites), see Ballestero, "Touching with Light."

4. In chapter 4 of *Program Earth,* I outline how citizenship could be understood ecologically through relations and practices, and also something that activates more-than-human relations and entities. This work draws on Rosi Braidotti's concept of the ecological citizen. Here, I extend this discussion of multiple citizens into an engagement with William James's notion of the multiverse, as well as the broader array of science and technology studies literature that works through ontologies of the multiple. See Gabrys, *Program Earth*; Braidotti, *Transpositions*; James, *A Pluralistic Universe*; and Mol, *The Body Multiple.*

5. Crawley, *Blackpentecostal Breath,* 3.

6. Crawley, 5. As Cornel West proposes in his analysis of pragmatism (drawing on Emerson), citizens form through the interactions of social life, which also give rise to the conditions of common experience. As this chapter further suggests, these interactions also involve more-than-humans as contributors to social interactions.

7. Kimmerer, *Braiding Sweetgrass,* 58.

8. Kimmerer, 58.

9. However, there are now multiple projects that have developed real-time and hybrid sensing and signaling technologies that enhance and convert plants' distributed sensing capabilities to detect pollution, explosives, and more. For instance, see "Plants Employed as Sensing Devices"; and Trafton, "Nanobionic Spinach Plants." In many ways, these projects are continuous with attempts to understand how plants sense and communicate. For example, see Karban, *Plant Sensing and Communication.*

10. For expanded discussions of more-than-human modes of witnessing, see Sheikh, "The Future of the Witness"; and Schuppli, *Material Witness.*

11. Van Haluwyn and van Herk, "Bioindication," 44.

12. Ferretti and Erhardt, "Key Issues in Designing Biomonitoring Programmes," 112.

13. Van Haluwyn and van Herk, "Bioindication," 56–58.

14. Van Haluwyn and van Herk, 40.

15. Kimmerer extensively discusses this making and remaking of earth and atmospheres. She writes: "We are all bound by a covenant of reciprocity: plant breath for animal breath, winter and summer, predator and prey, grass and fire, night and day, living and dying. Water knows this, clouds know this. Soil and rocks know they are dancing in a continuous giveaway of making, unmaking, and making again the earth." *Braiding Sweetgrass,* 383. While this chapter specifically investigates the relationships across plants and air, there are many other connections across water, other organisms, and soil that could be drawn out here if space permitted. Soil is now an especially dynamic area of research that points to further exploration across organisms, politics, technoscience, and ecologies. For example, see Puig de la Bellacasa, "Making Time for Soil"; and Lyons, *Vital Decomposition.*

16. As discussed throughout this book, world making as a concept and related set of terms now traverses multiple texts. While I draw on William James's use of this term, as well as ongoing pragmatist uses of it, I also follow this rich vein of thought into feminist technoscience, Indigenous theory and practice, and parallel literature. For the purposes of this chapter, I especially engage with Tsing, *The Mushroom at the End of the World*; Haraway, *When Species Meet*; Stengers, *Thinking with Whitehead*; and Kimmerer, *Braiding Sweetgrass*. Elaine Gan has also developed a Multispecies Worldbuilding Lab, which includes podcasts on this topic; see http://multispeciesworldbuilding.com/.

17. Tsing, *The Mushroom at the End of the World*, 21. As science and technology studies research would note, "nonliving" things also remake the world. This analysis necessarily extends to digital sensors and networks, among many other more-than-human entities.

18. Tsing, 22.

19. Simondon, *L'individuation*; Combes, *Gilbert Simondon*. See also Simondon, *Individuation*.

20. While Simondon's work examines these formations of entities and milieus, a parallel but different investigation into the "human" can be found in Sylvia Wynter's analysis of raced, classed, and sexed humans as misaligning with the universal human of colonialism or the Anthropocene. See Wynter and McKittrick, "Unparalleled Catastrophe for Our Species?"

21. Gilbert, Sapp, and Tauber, "A Symbiotic View of Life," 336.

22. Haraway begins her chapter "Tentacular Thinking: Anthropocene, Capitalocene, Chthulucene" with the epigraph "We Are All Lichens Now."

23. Kimmerer, *Braiding Sweetgrass*, 275.

24. Tsing, *The Mushroom at the End of the World*, 180.

25. Van Haluwyn and van Herk, "Bioindication," 48.

26. Van Haluwyn and van Herk, 50.

27. Van Haluwyn and van Herk, 53.

28. Gramaglia and Sampaio da Silva, "Researching Water Quality."

29. Kohn, *How Forests Think*.

30. More-than-human sensing as expressed through these multiplying points of view also demonstrates how "nature" is not a stable referent but rather a realm where diversity multiplies toward a "multinaturalism," where, as Eduardo Viveiros de Castro suggests, organisms might also be approached as persons and as having perspectives as persons. See Viveiros de Castro, *Cannibal Metaphysics*.

31. Such encounters with multiple environmental subjects and their worlds have further consequences for how "the ends of the world" are identified, averted, or addressed. See Danowski and Viveiros de Castro, *The Ends of the World*.

32. Stengers, *Cosmopolitics I*, 55. Cosmopolitics as developed by Stengers is a concept that indicates or asks how it might be possible to work toward different collective arrangements and processes that engage with humans and multiple more-than-humans, and thereby to transform the political scenes and environments of relevance within which problems come to matter. Plants, soil organisms, and fungi are viable actors that force different approaches to politics through cosmopolitical relations.

33. Tsing, *The Mushroom at the End of the World*, 254.

34. Simondon, *L'individuation*; and Combes, *Gilbert Simondon*. See also Simondon, *Individuation*.

35. Berlant, *The Queen of America*, 20.

36. TallBear, "Why Interspecies Thinking Needs Indigenous Standpoints." In a related way, Kimmerer notes that Indigenous knowledges allow for integral and democratic environmental engagements, where scientific analysis can tend to atomize its objects of analysis. See Kimmerer, *Braiding Sweetgrass*, 345.

37. Wenzel, "Reading Fanon Reading Nature," 189.

38. Simpson writes that the nation is a "place where we all live and work together" and "a web of connections to each other, to the plant nations, the animal nations, the rivers and lakes, the cosmos, and our neighboring Indigenous nations." See L. B. Simpson, *As We Have Always Done*, 8.

39. Rather than anthropomorphizing plants, such a move would even vegetalize citizens to allow for other encounters with environments as the conditions for being and becoming political subjects.

40. Resonating with Whitehead's approach to encountering everything as a subject, Kimmerer notes that "thinking about plants as persons, indeed, thinking about rocks as persons, forces us to shed our idea of the only pace that we live in is the human pace." Kimmerer, "The Intelligence of Plants." See also A. N. Whitehead, *Process and Reality*.

41. See City of London, "Barbican Low Emission Neighbourhood."

42. The work that more-than-humans undertake in processing carbon and pollutants is an ongoing area of research that I am developing further here. See also, Gabrys, "Plastic."

43. City in Bloom, *The Clean Air Gardens*. See also City of London Gardens; and City in Bloom, "Air Quality Challenge."

44. See Clean Air Gardens; and Friends of City Gardens.

45. Mapping for Change, "Science in the City: Barbican Citizen Science Project" and "Science in the City: Barbican Report."

46. This was also in the context of the City of London's development of a CityAir app to lower exposure to air pollution by suggesting walking routes away from heavy traffic. See https://www.cityoflondon.gov.uk/services/environmental-health/air-quality/cityair-app/.

47. For a discussion of community gardens and their contradictions, see McClintock, "Radical, Reformist, and Garden-Variety Neoliberal."

48. Museum of London, "How to Grow Your Own." The Phyto-Sensor Toolkit is also available to view and download on the Citizen Sense website at https://phyto-sensor-tool kit.citizensense.net. In the vein of the "how-to" guide that informs this book, there are also popular guides to purifying indoor air, including Wolverton, *How to Grow Fresh Air*.

49. The plants installed across the two demonstrator gardens included *Taxus baccata*, *Sorbaria sorbifolia*, *Hedera helix*, *Achillea millefolium*, *Aster* spp., *Sambucus racemosa*, *Symphoricarpos albus*, *Osmanthus delavayi*, *Salvia nemerosa*, *Lavandula* 'Platinum Blond', *Heuchera* spp., and *Euphorbia characias*. However, the Phyto-Sensor Toolkit includes a much more extensive list of air-quality plants, even if many of these are specific to UK and similar temperate environments.

50. For a resonant discussion of this theory and practice within Nishnaabeg intelligence, see L. B. Simpson, "Land as Pedagogy."

51. I include the Latin names here as a way to identify the organisms planted in the garden, but I also follow Kimmerer's suggestion not to let the scientific naming of plants become a strategy for closing down inquiry into established objects of analysis, but rather

to attend to the relationships and processes that are always unfolding and making worlds. See Kimmerer, "The Intelligence of Plants." For another example of how plants, relations, naming systems, and practices are co-constituted, see Siisip Geniusz, *Plants Have So Much to Give Us.*

52. City of London Corporation, "Air Quality Monitoring." The Beech Street air-quality monitor is one of five continuous monitoring locations in the City of London, which also has over one hundred nitrogen dioxide diffusion tubes. The City of London Corporation also provides air pollution alerts via the CityAir app and through airTEXT. See https://www.cityoflondon.gov.uk/services/environmental-health/air-quality/cityair-app/ and https://www.airtext.info.

53. Examples of how these multiple more-than-human interactions could unfold are discussed in Clarke et al., "More-than-Human Participation."

54. For an extensive range of references on these different pollution-absorbing and pollution-responding characteristics of plants, see Citizen Sense, "Phyto-Sensor Toolkit."

55. There is an extensive range of research on planting and power. While there is no space here to engage fully with this material, references that inform this discussion include Gray and Sheikh, "The Wretched Earth"; Tsing, *The Mushroom at the End of the World*; Haraway, *Staying with the Trouble*; Allewaert, *Ariel's Ecology*; Davis et al., "Anthropocene, Capitalocene . . . Plantationocene?"; Wynter and McKittrick, "Unparalleled Catastrophe for Our Species?"

56. Tuck et al., "Geotheorizing Black/Land," 55.

57. Tuck et al., 68. See also M. Smith, "Wildness."

58. Kimmerer, *Braiding Sweetgrass.*

59. Community gardening can be a way to activate "otherwise infrastructures," as LaDuke and Cowen discuss in "Beyond Wiindigo Infrastructure." For a parallel discussion in relation to the planetary, plantations, and the Anthropocene, see Yusoff, *A Billion Black Anthropocenes or None.*

60. Plumer and Popovich, "Decades of Racist Housing Policy."

61. Groundwork USA, "Climate Safe Neighborhoods."

62. The Phyto-Sensor Toolkit is available at https://phyto-sensor-toolkit.citizensense .net/.

63. Crawley, *Blackpentecostal Breath*, 11.

64. Crawley, 24, 79.

65. Crawley, 41.

66. Botanist John Gerard (along with others) is generally credited with having authored the *Herball*, a text that outlines medicinal uses of plants. See Gerard, *The Herball.* There has been a garden on this spot since at least 1555, and the Worshipful Company of Barber Surgeons (of which Gerard was Master of Company in 1607) created a Physic Garden to demonstrate plants involved with healing.

67. Crawley, *Blackpentecostal Breath*, 85.

68. A. N. Whitehead, *Process and Reality*, 78.

69. De la Cadena and Blaser explore the incommensurability of worlds and pluralism in their discussion of the uncommons. See de la Cadena and Blaser, *A World of Many Worlds.* In a somewhat different register than the uncommons, however, James's notion of pluralism engages with difference as requiring ongoing encounters, negotiations, and

adjustments to contingent worlds that are, at the same time, in contact with many other worlds.

70. For discussions of these different epistemologies and ontologies of citizen science, see Irwin, "Constructing the Scientific Citizen"; and Ottinger, "Buckets of Resistance."

71. See James, *Pragmatism and Other Writings*, 20 and 35; and James, *A Pluralistic Universe*, passim.

72. James, *Essays in Radical Empiricism*, 148.

73. James, *A Pluralistic Universe*.

74. In this use of the term "vegetating," Allewaert draws on William Bartram's travels through the United States, where he refers to plants and swamps more specifically as "vegetating" to indicate "the process by which a sentient force acts in an ecology that acts on and through it." See Allewaert, *Ariel's Ecology*, 34.

75. Richard J. Bernstein uses the term "engaged pluralism" to propose that James's pluralism did not lead to mere relativism. When we encounter other experiences and worlds, he notes, "This does *not* mean that when we make a serious effort to understand other points of view we will simply accept them or suspend our critical judgment. James's pluralism is not flabby or sentimental. It calls for a critical engagement with other points of view and with other visions. It is an engaged pluralism. Contrary to the picture of relativism that speaks of incommensurable frameworks and paradigms, James's pluralism demands that we reach out to the points of contact where we can critically engage with each other." See Bernstein, *The Pragmatic Turn*, 62.

76. Kimmerer, *Braiding Sweetgrass*, 373. As Simpson writes in a similar vein, these are struggles for particular ways of living, which span thought systems, bodies, temporalities, and practices. See L. B. Simpson, *As We Have Always Done*, 21. See also Celermajer et al., "Justice through a Multispecies Lens."

CONCLUSION

1. A. N. Whitehead, *Process and Reality*, 88.

2. Fanon, *A Dying Colonialism*; Crawley, *Blackpentecostal Breath*.

3. Gabrys, *Program Earth*, 162–63.

4. See also Sharpe, "The Weather"; and Gumbs, "That Transformative Dark Thing." Here, breathing is an accretion and accumulation, an ongoing struggle, as well as an opening into other possibilities for social, political, and worldly engagement.

5. James, *Essays in Radical Empiricism*, 153; see also Barry, "What Is an Environmental Problem?"

6. Consider, moreover, the networks of citizenship and noncitizenship that support the design, manufacture, and use of digital technologies: if people become citizens through sensors, what becomes of those humans and more-than-humans who are involved with and affected by the hazardous extraction of minerals for digital technologies, as well as the often-toxic conditions for the manufacture of these devices? The conditions of citizenship can signal toward the erasures and exclusions required to support particular designations and groupings of technopolitical subjects.

7. Or as Leanne Betasamosake Simpson has written, in the North American context for Indigenous peoples, "In situations in which sovereignties are nested and embedded, one proliferates at the other's expense." See L. B. Simpson, *As We Have Always Done*, 12.

8. While this resonates with pragmatist conceptions of democratic worlds and sub-jects, especially as articulated by Dewey, Glaude, and others, I am especially referring to Berlant's notion of "the affect of feeling political together," where "the attainment of that attunement produces a sense of shared worldness, apart from whatever aim or claim the listening public might later bring to a particular political world because of what they have heard." See Berlant, *Cruel Optimism*, 224. This also resonates with James's discus-sion of the formation of worlds and subjects, where the conditions of mutual experience involve meeting in common worlds (and perhaps, following de la Cadena, across uncom-mon worlds). As James suggests, mutuality and worlds are formed through meeting and experience. They are ongoing negotiations rather than absolute conditions. See James, *Essays in Radical Empiricism*, 79.

Bibliography

Adey, Peter. "Air/Atmospheres of the Megacity." *Theory, Culture & Society* 30, nos. 7–8 (2013): 291–308.

Agency for Toxic Substances and Disease Registry. "Health Consultation: Brooklyn Township $PM_{2.5}$." Atlanta: US Department of Health and Human Services, 2016.

Ahmed, Sara. *Living a Feminist Life*. Durham, N.C.: Duke University Press, 2017.

Air Quality Index. https://aqicn.org/city/delhi.

Air Sensors International Conference. 2018 program archive. https://asic2018.aqrc.ucda vis.edu/program.

Airviz. "Speck." https://www.specksensor.com/.

Alaimo, Stacy. *Bodily Natures: Science, Environment, and the Material Self*. Bloomington: Indiana University Press, 2010.

Allahyari, Morehshin, and Daniel Rourke, eds. *The 3D Additivist Cookbook*. Amsterdam: Institute of Network Cultures, 2016. http://additivism.org/cookbook.

Allewaert, Monique. *Ariel's Ecology: Plantations, Personhood, and Colonialism in the American Tropics*. Minneapolis: University of Minnesota Press, 2013.

Alliance for Aquatic Resource Monitoring. "Shale Gas Extraction: Volunteer Monitoring Manual." January 2017, version 3.3. http://www.dickinson.edu/download/downloads/ id/7012/shalegas manual 33.pdf.

Amrute, Sareeta, and Luis Felipe R. Murillo. "Introduction: Computing in/from the South." *Catalyst: Feminism, Theory, Technoscience* 6, no. 1 (2020): 1–23.

Anderson, Ben. "Affective Atmospheres." *Emotion, Space and Society* 2, no. 2 (2009): 77–81.

Arendt, Hannah. *The Human Condition*. 1958. Chicago: University of Chicago Press, 1998.

Arendt, Hannah. "Karl Jaspers: Citizen of the World?" In *Men in Dark Times*, 81–94. San Diego: Harcourt Brace, 1968.

Arendt, Hannah. *The Origins of Totalitarianism*. New York: Harcourt Brace, 1951.

Arnstein, Sherry. "A Ladder of Citizen Participation." *Journal of the American Planning Association* 35, no. 4 (July 1969): 216–24.

Asdal, Kirstin. "Enacting Things through Numbers: Taking Nature into Account/ing." *Geoforum* 39 (2008): 123–32.

Ashworth, Boone. "Inside Citizen, the App That Asks You to Report on the Crime Next Door." *Wired*, July 20, 2020. https://www.wired.com/story/citizen/.

Aunan, Kristin, Mette Halskov Hansen, and Shuxiao Wang. "Introduction: Air Pollution in China." *China Quarterly* 234 (2018): 279–98.

Austin, J. L. *How to Do Things with Words*. Cambridge, Mass.: Harvard University Press, 1975.

Backstrand, Karin. "Civic Science for Sustainability: Reframing the Role of Experts, Policy-Makers, and Citizens in Environmental Governance." *Global Environmental Politics* 3, no. 4 (November 2003): 24–41.

Balibar, Étienne. *Citizenship*. Cambridge, UK: Polity, 2015.

Balibar, Étienne, Barbara Cassin, and Sandra Laugier. "Praxis." In *Dictionary of Untranslatables: A Philosophical Lexicon*, edited by Barbara Cassin, translation edited by Emily Apter, Jacques Lezra, and Michael Wood, 820–32. Princeton, N.J.: Princeton University Press, 2014.

Ballestero, Andrea. "Touching with Light, or, How Texture Recasts the Sensing of Subterranean Water Worlds." *Science, Technology, & Human Values* 44, no. 5 (2019): 762–85.

Barad, Karen. *Meeting the Universe Halfway: Quantum Physics and the Entanglement of Matter and Meaning*. Durham, N.C.: Duke University Press, 2007.

Barry, Andrew. *Material Politics: Disputes along the Pipeline*. Oxford: Wiley-Blackwell, 2013.

Barry, Andrew. "What Is an Environmental Problem?" *Theory, Culture & Society* 38, no. 2 (2020): 93–117. https://doi.org/1177.0263276420958043.

Battle-Baptiste, Whitney, and Britt Rusert, eds. *W. E. B. Du Bois's Data Portraits*. Hudson, N.Y.: Princeton Architectural Press, 2018.

Bearak, Max. "Theresa May Criticized the Term 'Citizen of the World.' But Half the World Identifies That Way." *Washington Post*, October 5, 2016. https://www.washingtonpost.com/news/worldviews/wp/2016/10/05/theresa-may-criticized-the-term-citizen-of-the-world-but-half-the-world-identifies-that-way/.

Benjamin, Ruha, ed. *Captivating Technology: Race, Carceral Technoscience, and Liberatory Imagination in Everyday Life*. Durham, N.C.: Duke University Press, 2019.

Benjamin, Ruha. *People's Science: Bodies and Rights on the Stem Cell Frontier*. Stanford: Stanford University Press, 2013.

Benjamin, Ruha. *Race after Technology: Abolitionist Tools for the New Jim Code*. Cambridge, UK: Polity, 2019.

Benjamin, Ruha. "Racial Fictions, Biological Facts: Expanding the Sociological Imagination through Speculative Methods." *Catalyst: Feminism, Theory, Technoscience* 2, no. 2 (2016): 1–28.

Berlant, Lauren. "Citizenship." In *Keywords for American Cultural Studies*, edited by Bruce Burgett and Glenn Hendler, 37–42. New York: New York University Press, 2007.

Berlant, Lauren. "The Commons: Infrastructures for Troubling Times." *Environment and Planning D: Society and Space* 34, no. 3 (2016): 393–419.

Berlant, Lauren. *Cruel Optimism*. Durham, N.C.: Duke University Press, 2011.

Berlant, Lauren. *The Queen of America Goes to Washington City: Essays on Sex and Citizenship*. Durham, N.C.: Duke University Press, 1997.

Bernstein, Richard J. *The Pragmatic Turn*. Cambridge, UK: Polity, 2010.

Blackstone Chambers. "Inquest into the Death of Ella Adoo-Kissi-Debrah." December 17, 2020. https://www.blackstonechambers.com/news/inquest-death-ella-adoo-kissi-debrah.

Blas, Zach. *Gay Bombs: User's Manual*. Queer Technologies, 2008. http://www.zachblas .info/wp-content/uploads/2016/03/GB_users-manual_web-version.pdf.

Boehner, Kirsten, and Carl DiSalvo. "Data, Design, and Civics: An Exploratory Study of Civic Tech." In *Proceedings of the 2016 CHI Conference on Human Factors in Computing Systems* (2016): 2970–81.

Bolton, Matthew. *How to Resist: Turn Protest to Power*. London: Bloomsbury, 2017.

Borgonuovo, Valerio, and Silvia Franceschini. *Global Tools: 1973–1975*. İstanbul: SALT/ Garanti Kültür, 2015. http://saltonline.org/media/files/globaltools_scrd.pdf.

Boyd, David. "The Right to Breathe Clean Air." A/HRC/40/55 (2019). http://srenviron ment.org/report/the-right-to-breathe-clean-air-2019.

Boyle, Casey. "Pervasive Citizenship through #SenseCommons." *Rhetoric Society Quarterly* 46, no. 3 (2016): 269–83.

Braidotti, Rosi. *Transpositions: On Nomadic Ethics*. Cambridge, UK: Polity, 2006.

Breathe Easy Susquehanna Co. "Citizen Science Spurred Real Air Quality Monitoring in the PA Shalegas Fields." March 23, 2018. https://www.facebook.com/BreatheEasySusq /posts/citizen-science-spurred-real-air-quality-monitoring-in-the-pa-shalegas-fields-re /895271397320857/.

Briggle, Adam. *A Field Philosopher's Guide to Fracking*. New York: Liveright Publishing Corporation, 2015.

Bruno, Isabelle, Emmanuel Didier, and Tommaso Vitale. "Statactivism: Forms of Action between Disclosure and Affirmation." *Partecipazione e conflitto. The Open Journal of Sociopolitical Studies* 7, no. 2 (2014): 198–220.

Bruun Jensen, Casper, and Atsuro Morita. "Introduction: Infrastructures as Ontological Experiments." *Ethnos* 82, no. 4 (2016): 615–26.

Bucher, Taina. "Nothing to Disconnect From? Being Singular Plural in an Age of Machine Learning." *Media, Culture & Society* 42, no. 4 (2020): 610–17.

Bullard, Robert. *Dumping in Dixie: Race, Class and Environmental Quality*. Boulder, Colo.: Westview, 1990.

Bullard, Robert. "Solid Waste Sites and the Houston Black Community." *Sociological Inquiry* 53, nos. 2–3 (1983): 273–88.

Burke, Jeffrey A., Deborah Estrin, Mark Hansen, Andrew Parker, Nithya Ramanathan, Sasank Reddy, and Mani B. Srivastava. "Participatory Sensing." In *Proceedings of the World Sensor Web Workshop*. Boulder, Colo.: ACM SENSYS, 2006. https://escholarship .org/uc/item/19h777qd#page-1.

Butler, Judith. *Bodies That Matter: On the Discursive Limits of "Sex."* New York: Routledge, 1993.

Butler, Judith. *Excitable Speech: A Politics of the Performative*. New York: Routledge, 1997.

Buzzelli, Michael, Michael Jerrett, Richard Burnett, and Norm Finklestein. "Spatiotemporal Perspectives on Air Pollution and Environmental Justice in Hamilton, Canada, 1985–1996." *Annals of the Association of American Geographers* 93, no. 3 (2003): 557–73.

Campt, Tina. "John Akomfrah, in Conversation with Tina Campt, Ekow Eshun, Saidiya Hartman." Lisson Studio. June 18, 2020. https://www.lissongallery.com/studio/john -akomfrah-tina-campt-saidiya-hartman.

Care for Air. http://www.careforair.org/.

Carrington, Damian. "Covid-19 Impact on Ethnic Minorities Linked to Housing and Air Pollution." *The Guardian*, July 19, 2020. https://www.theguardian.com/world/2020/ jul/19/covid-19-impact-on-ethnic-minorities-linked-to-housing-and-air-pollution.

Celermajer, Danielle, Sria Chatterjee, Alasdair Cochrane, Stefanie Fishel, Astrida Nei-
manis, Anne O'Brien, Susan Reid, Krithika Srinivasan, David Schlosberg, and Anik
Waldow. "Justice through a Multispecies Lens." *Contemporary Political Theory* 19
(2020): 475–512.

Center for Urban Pedagogy. "Making Policy Public." http://welcometocup.org/Projects/
MakingPolicyPublic.

Césaire, Aimé. *Discourse on Colonialism.* Translated by Joan Pinkham. 1972. New York:
Monthly Review Press, 2000.

Chariton, Jordan. Interview with Dean Deadman. "Drone Pilots Exposing Oil Police Violence." The Young Turks, November 16, 2016. https://www.youtube.com/watch?v=R
5QW3H_0FiM.

Cheah, Pheng. *What Is a World? On Postcolonial Literature as World Literature.* Durham,
N.C.: Duke University Press, 2016.

Chen, Mel Y. *Animacies: Biopolitics, Racial Mattering, and Queer Affect.* Durham, N.C.:
Duke University Press, 2012.

Chilvers, Jason, and Matthew Kearnes, eds. *Remaking Participation: Science, Environment,
and Emergent Publics.* London: Routledge, 2016.

Choy, Tim. *Ecologies of Comparison: An Ethnography of Endangerment in Hong Kong.* Durham, N.C.: Duke University Press, 2011.

Choy, Timothy, and Jerry Zee. "Condition—Suspension." *Cultural Anthropology* 30, no. 2
(2015): 210–23.

Chrisler, Matt, Jaskiran Dhillon, and Audra Simpson. "The Standing Rock Syllabus Project." October 21, 2016. http://www.publicseminar.org/2016/10/nodapl-syllabus-proj
ect/#.WEiq1KIr1xG.

Chude-Sokei, Louis. *The Sound of Culture: Diaspora and Black Technopoetics.* Middletown,
Conn.: Wesleyan University Press, 2015.

Cifor, Marika, Patricia Garcia, T. L. Cowan, Jasmine Rault, Tonia Sutherland, Anita Say
Chan, Jennifer Rode, Anna Lauren Hoffmann, Niloufar Salehi, and Lisa Nakamura.
"Feminist Data Manifest-No." 2019. https://www.manifestno.com/.

Citizen Sense. "Citizen Sense Frackbox." https://citizensense.net/kits/frackbox-hardware.

Citizen Sense. "Citizen Sense Kit." https://citizensense.net/kits/citizensense-kit.

Citizen Sense. "Citizen Sense Monitoring Events in Pennsylvania." October 24, 2014.
https://citizensense.net/citizen-sense-monitoring-kit-pennsylvania.

Citizen Sense. "Covid Data Stories." December 30, 2020. https://citizensense.net/data
-stories-covid.

Citizen Sense. "Deptford Data Stories." November 14, 2017. https://citizensense.net/data
-stories-deptford.

Citizen Sense. "Pennsylvania Data Stories." April 19, 2016. https://citizensense.net/data
-stories-pa.

Citizen Sense. "Phyto-Sensor Toolkit." https://phyto-sensor-toolkit.citizensense.net/.

Citizen Sense. "Pollution Sensing Videos." https://citizensense.net/videos/pollution-sens
ing-videos/.

Citizen Sense. "Urban Sensing Using the Dustbox (Workshop and Walk)." October 2016.
https://vimeo.com/203283403.

CityAir app. https://www.cityoflondon.gov.uk/services/environmental-health/air-quality/
cityair-app/.

City in Bloom. "Air Quality Challenge." http://www.cityinbloom.org/2017.html.

City in Bloom. *The Clean Air Gardens*. City of London, 2017.

City of London. "Barbican Low Emission Neighbourhood." http://democracy.cityoflondon .gov.uk/documents/s69460/LEN%20bid%20-%20background%20paper.pdf.

City of London Corporation. "Air Quality Monitoring." August 4, 2020. https://www.cityof london.gov.uk/services/environmental-health/air-quality/air quality-monitoring; https:// www.cityoflondon.gov.uk/services/environmental-health/air-quality/cityair-app/; https: //www.airtext.info.

City of London Gardens. https://twitter.com/CoLCityGdns.

Clark, Anna. *The Poisoned City*: *Flint's Water and the American Urban Tragedy*. New York: Metropolitan Books, 2018.

Clarke, Rachel, Sara Heitlinger, Ann Light, Laura Forlano, Marcus Foth, and Carl DiSalvo. "More-than-Human Participation: Design for Sustainable Smart City Futures." *Interactions* 26, no. 3 (2019): 60–63.

Clean Air Gardens. http://cleanairgardens.blogspot.co.uk/.

ClientEarth Communications. "ClientEarth Launches New UK Air Pollution Legal Action." November 7, 2017. https://www.clientearth.org/latest/latest-updates/news/clientearth -launches-new-uk-air-pollution-legal-action/.

Climate and Clean Air Coalition. "Satellite Data Reveals Extreme Methane Emissions from US Permian Oil & Gas Operations." April 23, 2020. https://ccacoalition.org/en/ news/satellite-data-reveals-extreme-methane-emissions-us-permian-oil-gas-operation.

Climate and Clean Air Coalition. "USA Offshore Oil and Gas Platforms Release More Methane than Previously Estimated." April 14, 2020. https://ccacoalition.org/en/ news/usa-offshore-oil-and-gas-platforms-release-more-methane-previously-estimated.

Cocco, Federica, and Alan Smith. "Race and America: Why Data Matters." *Financial Times Magazine*, July 23, 2020. https://www.ft.com/content/156f770a-1d77-4f6b-8616-192fb 58e3735.

Cohen, Ron C. "Challenges and Benefits of Backyard Science." *Sensing Change*. Philadelphia: Science History Institute. July 18, 2013. https://www.sciencehistory.org/ronald -c-cohen.

Colaneri, Katie. "Gas Drilling Draws Citizen Scientists to the Field." *State Impact: Pennsylvania (NPR)*, December 19, 2014. https://stateimpact.npr.org/pennsylvania/2014/.

Combes, Muriel. *Gilbert Simondon and the Philosophy of the Transindividual*. Translated by Thomas Lamarre. 1999. Cambridge, Mass.: MIT Press, 2013.

Corburn, Jason. *Street Science: Community Knowledge and Environmental Health Justice*. Cambridge, Mass.: MIT Press, 2005.

Corporate Watch. "Tidemill: Factsheet on the Battle for Deptford." November 12, 2018. https://corporatewatch.org/tidemill-development-factsheet-on-the-battle-for-deptford/.

Couldry, Nick, and Alison Powell. "Big Data from the Bottom Up." *Big Data & Society* 1, no. 2 (2014): 2053951714539277.

Cox, Sarah. "Citizens' Digital Monitoring Project Shows Evidence of Air Pollution near US Fracking Sites." Goldsmiths, University of London, April 19, 2016. https://www .gold.ac.uk/news/citizen-sense-data-stories.

Crawley, Ashon T. *Blackpentecostal Breath: The Aesthetics of Possibility*. New York: Fordham University Press, 2017.

Create Lab. "CATTfish." http://www.cmucreatelab.org/projects/Water_Quality_Monitor ing/pages/CATTfish; https://www.cattfish.com/.

Create Lab. "Speck." https://www.cmucreatelab.org/projects/Speck.

Crosswhatfields? "No. 1 Creekside—Lewisham Passes Another Bad Scheme without Any Scrutiny." April 20, 2019. http://crossfields.blogspot.com/2019/04/no1-creekside-lew isham-passes-another.html.

Cuffe, Grainne. "Lewisham Paying Back Debt to County Enforcement Bailiff." *This Is Local London*, June 12, 2019. https://www.thisislocallondon.co.uk/news/18513337.lew isham-paying-back-debt-county-enforcement-bailiff/.

Currie, Morgan, Britt S. Paris, Irene Pasquetto, and Jennifer Pierre. "The Conundrum of Police Officer–Involved Homicides: Counter-Data in Los Angeles County." *Big Data & Society* 3, no. 2 (2016): 1–14.

Cvetkovich, Ann. *Depression: A Public Feeling*. Durham, N.C.: Duke University Press, 2012.

da Costa, Beatriz, and Kavita Philip, eds. *Tactical Biopolitics: Art, Activism, and Technoscience*. Cambridge, Mass.: MIT Press, 2008.

Daniek, Michel. *Do It Yourself 12 Volt Solar Power*. East Meon, UK: Permanent Publications, 2007.

Danowski, Déborah, and Eduardo Viveiros de Castro. *The Ends of the World*. Translated by Rodrigo Nunes. 2014. Cambridge, UK: Polity, 2017.

Dányi, Endre. "The Politics of 'How.'" *Does STS Have Problems?* October 4, 2016. https:// stsproblems.wordpress.com/2016/10/04/the-politics-of-how.

Das, Pamela, and Richard Horton. "Pollution, Health, and the Planet: Time for Decisive Action." *The Lancet*, October 19, 2017. https://doi.org/10.1016/S0140-6736(17)32588 -6/.

Daston, Lorraine, and Peter Galison. *Objectivity*. Brooklyn: Zone Books, 2007.

Data for Black Lives. https://d4bl.org/.

Datta, Ayona. "The Digital Turn in Postcolonial Urbanism: Smart Citizenship in the Making of India's 100 Smart Cities." *Transactions of the Institute of British Geographers* 43, no. 3 (2018): 405–19.

Davenport, Coral. "Trump Eliminates Major Methane Rule, Even as Leaks Are Worsening." *New York Times*, August 20, 2020. https://www.nytimes.com/2020/08/13/cli mate/trump-methane.html.

Davis, Janae, Alex A. Moulton, Levi Van Sant, and Brian Williams. "Anthropocene, Capitalocene, . . . Plantationocene? A Manifesto for Ecological Justice in an Age of Global Crises." *Geography Compass* 13, no. 5 (2019): e12438.

DeGooyer, Stephanie, Alastair Hunt, Lida Maxwell, and Samuel Moyn. *The Right to Have Rights*. London: Verso, 2018.

de la Cadena, Marisol. "Indigenous Cosmopolitics in the Andes: Conceptual Reflections beyond 'Politics.'" *Cultural Anthropology* 25, no. 2 (2010): 334–70.

de la Cadena, Marisol, and Mario Blaser, eds. *A World of Many Worlds*. Durham, N.C.: Duke University Press, 2018.

de Laet, Marianne, and Annemarie Mol. "The Zimbabwe Bush Pump: Mechanics of a Fluid Technology." *Social Studies of Science* 30, no. 2 (2000): 225–63.

Delgado, Ana, and Blanca Callén. "Do-It-Yourself Biology and Electronic Waste Hacking: A Politics of Demonstration in Precarious Times." *Public Understanding of Science* 26, no. 2 (2017): 179–94.

Dencik, Lina, Arne Hintz, and Jonathan Cable. "Towards Data Justice? The Ambiguity of Anti-surveillance Resistance in Political Activism." *Big Data & Society* 3, no. 2 (2016): 1–12.

Denzin, Norman K., Yvonna S. Lincoln, and Linda Tuhiwai Smith, eds. *Handbook of Critical and Indigenous Methodologies*. Thousand Oaks, Calif.: Sage, 2008.

Department of Environmental Protection Bureau of Oil and Gas Management. "Chemicals Used by Hydraulic Fracturing Companies in Pennsylvania for Surface and Hydraulic Fracturing Activities." http://files.dep.state.pa.us/OilGas/BOGM/BOGM PortalFiles/MarcellusShale/Frac%20list%206-30-2010.pdf.

Department of Environment, Food, and Rural Affairs. "Graph of Hourly Measurements for London Marylebone Road." https://uk-air.defra.gov.uk/data-plot?site_id=MY1&days=7.

Department of Environment, Food, and Rural Affairs. "Interactive Monitoring Networks Map." https://uk-air.defra.gov.uk/interactive-map.

Department of Environment, Food, and Rural Affairs. "Site Information for London Marylebone Road (UKA00315)." https://uk-air.defra.gov.uk/networks/site-info?site_id=MY1.

Deptford Folk. https://www.deptfordfolk.org/.

Deptford Parks. https://deptfordparks.commonplace.is/.

Descartes, René. *Discourse on Method and Related Writings*. Translated by Desmond M. Clarke. 1637. London: Penguin, 2003.

Detroit Community Technology Project. https://detroitcommunitytech.org/.

Dewey, John. "The Development of American Pragmatism." In *John Dewey: The Later Works, 1925–1953*, vol. 2, *1925–1927*, 3–21. Carbondale: Southern Illinois University Press, 2008.

Dewey, John. *Human Nature and Conduct: An Introduction to Social Psychology*. Amherst, UK: Prometheus Books, 2002.

Dewey, John. *John Dewey: The Later Works, 1925–1953*. Vols. 1–4. Carbondale: Southern Illinois University Press, 2008.

Dewey, John. *Logic: The Theory of Inquiry*. New York: Henry Holt, 1938.

Dewey, John. "The Naturalization of Intelligence." In *John Dewey: The Later Works*, vol. 4, *1929*, 156–77. Carbondale: Southern Illinois University Press, 2008.

Dewey, John. *The Public and Its Problems: An Essay in Political Inquiry*. Edited and with an introduction by Melvin L. Rogers. University Park: Pennsylvania State University Press, 2012.

Dewey, John. *The Quest for Certainty. John Dewey: The Later Works, 1925–1953*, vol. 4, *1929*. Carbondale: Southern Illinois University Press, 2008.

Dhanjani, Nitesh. *Abusing the Internet of Things: Blackouts, Freakouts, and Stakeouts*. Beijing: O'Reilly, 2015.

Dhillon, Carla May. "Using Citizen Science in Environmental Justice: Participation and Decision-Making in a Southern California Waste Facility Siting Conflict." *Local Environment* 22, no. 12 (2017): 1479–96.

Dhillon, Jaskiran. *Prairie Rising: Indigenous Youth, Decolonization, and the Politics of Intervention*. Toronto: University of Toronto Press, 2017.

Dhillon, Jaskiran, and Nick Estes, eds. *Standing with Standing Rock: Voices from the #NoDAPL Movement*. Minneapolis: University of Minnesota Press, 2019.

Dillon, Lindsey, and Julie Sze. "Police Power and Particulate Matters: Environmental Justice and the Spatialities of In/Securities in US Cities." *English Language Notes* 54, no. 2 (Fall/Winter 2016): 13–23.

DiSalvo, Carl, Marti Louw, David Holstius, Illah Nourbakhsh, and Ayça Akin. "Toward a Public Rhetoric through Participatory Design: Critical Engagements and Creative Expression in the Neighborhood Networks Project." *Design Issues* 28, no. 3 (2012): 48–61.

Dockery, D. W., C. A. Pope, X. Xu, J. D. Spengler, J. H. Ware, M. E. Fay, B. G. Ferris Jr., and F. E. Speizer. "An Association between Air Pollution and Mortality in Six US Cities." *New England Journal of Medicine* 329, no. 24 (1993): 1753–59.

Don't Dump on Deptford's Heart. Don't Dump on Deptford's Heart Google map. https://www.google.com/maps/d/viewer?hl=en_US&mid=13Xl4bTJ-wYv8LQEfjNqw_vpaot8&ll=51.48235499179066%2C-0.029080000000021755&z=14.

Don't Dump on Deptford's Heart. "Help Us Combat Thames Tunnel Pollution." November 29, 2013. https://dontdumpondeptfordsheart.wordpress.com/2013/11/29/help-us-combat-thames-tunnel-pollution/.

Dourish, Paul, and W. Keith Edwards. "A Tale of Two Toolkits: Relating Infrastructure and Use in Flexible CSCW Toolkits." *Computer Supported Cooperative Work* 9 (2000): 33–51.

Drone2bewild. "Drone Footage of Dakota Access Pipeline Approaching Missouri River." Interviewed by Paulette Moore. November 1, 2016. https://vimeo.com/189876726.

Duarte, Marisa Elena. *Network Sovereignty: Building the Internet across Indian Country.* Seattle: University of Washington Press, 2017.

Earthworks. "Gas Patch Roulette: How Shale Gas Development Risks Public Health in Pennsylvania." Earthworks Oil and Gas Accountability Project. October 2012. https://earthworks.org/publications/gas_patch_roulette_full_report.

Edwards, Pete. "Pigeon Air Patrol: A Realistic Way of Monitoring Pollution or Cooing over Unproved Science?" *The Conversation*, March 17, 2016. https://theconversation.com/pigeon-air-patrol-a-realistic-way-of-monitoring-pollution-or-cooing-over-unproved-science-56315.

El Recetario [The cookbook]. http://el-recetario.net/.

The Ella Roberta Family Foundation. http://ellaroberta.org.

The Ella Roberta Family Foundation. "Every Breath Matters." August 11, 2019. http://ellaroberta.org/aiovg_videos/this-is-ella/.

Elliott, Kevin C., Robert Montgomery, David B. Resnick, and Robert Goodwin. "Drone Use for Environmental Research." *IEEE Geoscience and Remote Sensing Magazine* 7, no. 1 (2019): 106–11.

Elwood, Sarah. "Volunteered Geographic Information: Future Research Directions Motivated by Critical, Participatory, and Feminist GIS." *GeoJournal* 72, nos. 3–4 (2008): 173–83.

End Child Poverty. http://www.endchildpoverty.org.uk/local-child-poverty-data-2014-15-20 19-20.

English, P. B., M. J. Richardson, and Catalina Garzón-Galvis. "From Crowdsourcing to Extreme Citizen Science: Participatory Research for Environmental Health." *Annual Review of Public Health* 39 (2018): 335–50.

Environmental Defense Fund. "With This Technology, Methane Leaks Never Out of Sight, Out of Mind." December 11, 2014. https://www.edf.org/blog/2014/12/11/tech nology-methane-leaks-never-out-sight-out-mind.

Estes, Nick. *Our History Is the Future: Standing Rock versus the Dakota Access Pipeline, and the Long Tradition of Indigenous Resistance*. London: Verso, 2019.

Eubanks, Virginia. "Double-Bound: Putting the Power Back into Participatory Research." *Frontiers: A Journal of Women Studies* 30, no. 1 (2009): 107–37.

European Commission. Directive 2008/50/EC of the European Parliament and of the Council of 21 May 2008 on Ambient Air Quality and Cleaner Air for Europe. *Official Journal of the European Union* (November 6, 2008): L 152/1.

Evelyn, John. *Fumifugium*. 1661. Exeter: The Rota, 1976.

Fanon, Frantz. *Black Skin, White Masks*. Translated by Charles L. Markmann. New York: Grove Press, 1967.

Fanon, Frantz. *A Dying Colonialism*. Translated by Haakon Chevalier. New York: Grove Press, 1965.

Feel Tank Chicago. "The Feel Kit." http://feeltankchicago.net/.

Feenberg, Andrew. *Questioning Technology*. Abingdon, UK: Routledge, 1999.

Felt, Ulrike, and Maximilian Fochler. "The Bottom-up Meanings of the Concept of Public Participation in Science and Technology." *Science and Public Policy* 35, no. 7 (2008): 489–99.

Fennell, Catherine. "Are We All Flint?" *Limn* 7: "Public Infrastructures/Infrastructural Publics," edited by Stephen J. Collier, James Christopher Mizes, and Antina von Schnitzler. July 2016. https://limn.it/articles/are-we-all-flint/.

Ferretti, Marco, and Walter Erhardt. "Key Issues in Designing Biomonitoring Programmes: Monitoring Scenarios, Sampling Strategies, and Quality Assurance." In *Monitoring with Lichens—Monitoring Lichens*, edited by Pier Luigi Nimis, Christoph Scheidegger, and Patricia A. Wolseley, 111–39. Amsterdam: Kluwer Academic Publishers, 2002.

Finan, Frank. https://www.youtube.com/channel/UC7Eph33czawYR2ZKZrexS0Q.

Fish, Adam. "Crash Theory: Entrapments of Conservation Drones and Endangered Megafauna." *Science, Technology, & Human Values* 46, no. 2 (2020): 425–51. https://doi.org/10.1177%2F0162243920920356.

"Flood Network." https://flood.network/.

Fortun, Kim. "From Latour to Late Industrialism." *HAU: Journal of Ethnographic Theory* 4, no. 1 (2014): 309–29.

Fortun, Kim, and Michael Fortun. "Scientific Imaginaries and Ethical Plateaus in Contemporary US Toxicology." *American Anthropologist* 107, no. 1 (2005): 43–54.

Fortun, Kim, Lindsay Poirier, Alli Morgan, Brandon Costelloe-Kuehn, and Mike Fortun. "Pushback: Critical Data Designers and Pollution Politics." *Big Data & Society* 3, no. 2 (2016): 1–14.

Foxcroft, Vicky (Lewisham, Deptford, Lab). "Business of the House." *House of Commons Hansard* 632 (December 7, 2017). https://hansard.parliament.uk/Commons/2017-12-07/debates/A0CAD9B5-6A6C-415E-9D7F-BED111E4D4AC/BusinessOfTheHouse#contribution-C9B74F07-D33B-4A59-817D-4A2C864BE3AB.

Fractracker Alliance. https://www.fractracker.org.

Fractracker Alliance. "Pennsylvania Shale Viewer." March 5, 2020. https://www.frac tracker.org/map/us/pennsylvania/pa-shale-viewer.

Frickel, Scott, Sahra Gibbon, Jeff Howard, Joanna Kempner, Gwen Ottinger, and David J. Hess. "Undone Science: Charting Social Movement and Civil Society Challenges to Research Agenda Setting." *Science, Technology, & Human Values* 35, no. 4 (2010): 444–73.

Friends of City Gardens. https://friendsofcitygardens.org.uk/.

Gabrys, Jennifer. "Air Walk: Monitoring Pollution and Experimenting with Speculative Forms of Participation." In *Walking through Social Research*, edited by Charlotte Bates and Alex Rhys-Taylor, 144–61. London: Routledge, 2017.

Gabrys, Jennifer. "Atmospheres of Communication." In *Circulation and the City: Essays on Urban Culture*, edited by Alexandra Boutros and Will Straw, 46–59. Montreal: McGill University Press, 2010.

Gabrys, Jennifer. "Becoming Urban: Sitework from a Moss-Eye View." *Environment and Planning A* 44 (2012): 2922–39.

Gabrys, Jennifer. "Citizen Sensing, Environmental Monitoring, and 'Media' as Practice in the Making." In *The Routledge Companion to Media Studies and Digital Humanities*, edited by Jentery Sayers, 503–10. New York: Routledge, 2018.

Gabrys, Jennifer. "Citizen Sensing: Recasting Digital Ontologies through Proliferating Practices." Theorizing the Contemporary, Cultural Anthropology website. March 24, 2016. https://culanth.org/fieldsights/citizen-sensing-recasting-digital-ontologies -through-proliferating-practices.

Gabrys, Jennifer. "A Cosmopolitics of Energy: Diverging Materialities and Hesitating Practices." *Environment and Planning A* 46, no. 9 (2014): 2095–109.

Gabrys, Jennifer. *Digital Rubbish: A Natural History of Electronics*. Ann Arbor: University of Michigan Press, 2011.

Gabrys, Jennifer. "For the World, Not of the World." *Metascience* 23, no. 3 (2014): 513–16.

Gabrys, Jennifer. "Planetary Health in Practice: Sensing Air Pollution and Transforming Urban Environments." *Humanities and Social Sciences Communications* 7, no. 1 (2020): 1–11.

Gabrys, Jennifer. "Plastic and the Work of the Biodegradable." In *Accumulation: The Material Politics of Plastic*, edited by Jennifer Gabrys, Gay Hawkins, and Mike Michael, 208–27. London: Routledge, 2013.

Gabrys, Jennifer. *Program Earth: Environmental Sensing Technology and the Making of a Computational Planet*. Minneapolis: University of Minnesota Press, 2016.

Gabrys, Jennifer. "Programming Environments: Environmentality and Citizen Sensing in the Smart City." *Environment and Planning D: Society and Space* 32, no. 1 (2014): 30–48.

Gabrys, Jennifer. "Sensing Air and Creaturing Data." In Jennifer Gabrys, *Program Earth: Environmental Sensing Technology and the Making of a Computational Planet*, 157–81. Minneapolis: University of Minnesota Press, 2016.

Gabrys, Jennifer. "Sensing a Planet in Crisis." *Media+Environment* 1, no. 1 (2019). https:// doi.org/10.1525/001c.10036.

Gabrys, Jennifer, Helen Pritchard, and Benjamin Barratt. "Just Good Enough Data: Figuring Data Citizenships through Air Pollution Sensing and Data Stories." *Big Data & Society* 3, no. 2 (2016): 1–14.

Gabrys Jennifer, and Kathryn Yusoff. "Arts, Sciences, and Climate Change: Practices and Politics at the Threshold." *Science as Culture* 21, no. 1 (2012): 1–24.

Gad, Christopher, and Casper Bruun Jensen. "The Promises of Practice." *Sociological Review* 62, no. 4 (2014): 698–718.

Gan, Elaine. Multispecies Worldbuilding Lab. http://multispeciesworldbuilding.com/.

Gane, Nicholas, and Donna Haraway. "When We Have Never Been Human, What Is to Be Done? Interview with Donna Haraway." *Theory, Culture, and Society* 23, nos. 7–8 (2006): 135–58.

Geiling, Natasha. "Trump Administration Officially Scraps Obama-Era Rules for Fracking on Federal Land." *ThinkProgress*, January 2, 2018. https://thinkprogress.org/trump-administration-repeals-obama-fracking-rules-f4595696dff7/.

Gerard, John. *The Herball, or Generall Historie of Plants*. London: John Norton, 1597.

Ghertner, D. Asher. "Airpocalypse: Distributions of Life amidst Delhi's Polluted Airs." *Public Culture* 32, no. 1 (2020): 133–62.

Gilbert, Scott F., Jan Sapp, and Alfred I. Tauber. "A Symbiotic View of Life: We Have Never Been Individuals." *Quarterly Review of Biology* 87, no. 4 (2012): 325–41.

Gill, Natalie. "Caring for Clean Streets: Policies as World-Making Practices." *Sociological Review* 65, issue 2_suppl (2017): 71–88.

Gill, Natalie, Vicky Singleton, and Claire Waterton. "The Politics of Policy Practices." *Sociological Review* 65, issue 2_suppl (2017): 3–19.

Glaude, Eddie S. *In a Shade of Blue: Pragmatism and the Politics of Black America*. Chicago: University of Chicago Press, 2008.

Global Forest Watch. https://www.globalforestwatch.org.

Goodchild, Michael F. "Citizens as Sensors: The World of Volunteered Geography." *GeoJournal* 69 (2007): 211–21.

Górska, Magdalena. "Breathing Matters: Feminist Intersectional Politics of Vulnerability." *Linköping Studies in Arts and Science*, no. 683. Linköping University, 2016.

Graham, Stephen. "Life Support: The Political Ecology of Urban Air." *City* 19, nos. 2–3 (2015): 192–215.

Gramaglia, Christelle, and François Mélard. "Looking for the Cosmopolitical Fish: Monitoring Marine Pollution with Anglers and Congers in the Gulf of Fos, Southern France." *Science, Technology, & Human Values* 44, no. 5 (2019): 814–42.

Gramaglia, Christelle, and Delaine Sampaio da Silva. "Researching Water Quality with Non-humans: An ANT Account." In *Agency without Actors? New Approaches to Collective Action*, edited by Jan-Hendrik Passoth, Birgit Peuker, and Michael Schillmeier, 184–85. Abingdon, UK: Routledge, 2012.

Gray, Ros, and Shela Sheikh. "The Wretched Earth: Botanical Conflicts and Artistic Interventions." *Third Text* 32, nos. 2–3 (2018): 163–75.

Green, Lesley. "Fracking the Karoo." In *Rock | Water | Life: Ecology and Humanities for a Decolonial South Africa*, 60–76. Durham, N.C.: Duke University Press, 2020.

Gregory, Judith, and Geoffrey C. Bowker. "The Data Citizen, the Quantified Self, and Personal Genomics." In *Quantified: Biosensing Technologies in Everyday Life*, edited by Dawn Nafus, 211–26. Cambridge, Mass.: MIT Press, 2016.

Griffith Spears, Ellen. *Baptized in PCBs: Race, Pollution, and Justice in an All-American Town*. Chapel Hill: University of North Carolina Press, 2014.

Grigg, Jonathan. "Where Do Inhaled Fossil Fuel–Derived Particles Go?" *American Journal of Respiratory and Critical Care Medicine* 196, no. 7 (2017): 804–6.

Griswold, Eliza. *Amity and Prosperity: One Family and the Fracturing of America*. New York: Farrar, Straus, and Giroux, 2018.

Griswold, Eliza. "The Fracturing of Pennsylvania." *New York Times*, November 17, 2011. http://www.nytimes.com/2011/11/20/magazine/fracking-amwell-township.html.

Groundwork USA. "Climate Safe Neighborhoods." https://groundworkusa.org/climate-safe-neighborhoods/; https://gwmke.maps.arcgis.com/apps/Cascade/index.html?appid=9b784d9e79324d1f97210b25afe1b91d.

Gumbs, Alexis Pauline. *M Archive: After the End of the World*. Durham, N.C.: Duke University Press, 2018.

Gumbs, Alexis Pauline. "That Transformative Dark Thing." *The New Inquiry*, May 19, 2015. https://thenewinquiry.com/that-transformative-dark-thing/.

Gutiérrez, Miren. *Data Activism and Social Change*. London: Palgrave Macmillan, 2018.

Hacking, Ian. "Canguilhem amid the Cyborgs." *Economy and Society* 27, nos. 2–3 (1998): 202–16.

Hacking, Ian. *Representing and Intervening: Introductory Topics in the Philosophy of Natural Science*. Cambridge: Cambridge University Press, 1983.

Hancock, Sam. "Khan Calls Lewisham Emissions 'Health Crisis' as Council Cuts Air Quality Funding." *East London Lines*, November 23, 2018. https://www.eastlondonlines.co.uk/2018/11/khan-calls-lewisham-emission-health-crisis-as-council-cuts-air-quality-funding.

Hanna-Attisha, Mona. *What the Eyes Don't See: A Story of Crisis, Resistance, and Hope in an American City*. London: One World, 2018.

Haraway, Donna. *Modest_Witness@Second_Millennium.FemaleMan©_Meets_OncoMouse™*. New York: Routledge, 1997.

Haraway, Donna. *Simians, Cyborgs, and Women: The Reinvention of Nature*. New York: Routledge, 1991.

Haraway, Donna. *Staying with the Trouble: Making Kin in the Chthulucene*. Durham, N.C.: Duke University Press, 2016.

Haraway, Donna. "Tentacular Thinking: Anthropocene, Capitolocene, Plantationocene, Chthulucene." In *Staying with the Trouble: Making Kin in the Chthulucene*, 30–57. Durham, N.C.: Duke University Press, 2016.

Haraway, Donna. *When Species Meet*. Minneapolis: University of Minnesota Press, 2008.

Hawkins, Gay. "Governing Litter: Habits, Infrastructures, Atmospheres." In *Assembling and Governing Habits*, edited by Tony Bennett, Ben Dibley, Gay Hawkins, and Greg Noble, 81–96. Abingdon, UK: Routledge, 2021.

Hecht, Gabrielle. "Air in a Time of Oil." *Los Angeles Review of Books*, January 21, 2019. https://blog.lareviewofbooks.org/provocations/air-time-oil.

Helmreich, Stefan. "Reading a Wave Buoy." *Science, Technology, & Human Values* 44, no. 5 (September 2019): 737–61.

Hemmi, Akiko, and Ian Graham, "Hacker Science versus Closed Science: Building Environmental Monitoring Infrastructure." *Information, Communication & Society* 17, no. 7 (2014): 830–42.

Hickey, Amber, ed. *A Guidebook of Alternative Nows*. The Journal of Aesthetics and Protest Press, 2012. https://joaap.org/press/alternativenows_hickey.htm.

Higgins, Hannah. *Fluxus Experience*. Berkeley: University of California Press, 2002.

Hintz, Arne, Lina Dencik, and Karin Wahl-Jorgensen. *Digital Citizenship in a Datafied Society*. Cambridge, UK: Polity, 2018.

Holgate, Stephen T. "Every Breath We Take: The Lifelong Impact of Air Pollution—A Call for Action." *Clinical Medicine* 17, no. 1 (2017): 8–12.

Houston, Donna. "Environmental Justice Storytelling: Angels and Isotopes at Yucca Mountain, Nevada." *Antipode* 45, no. 2 (2013): 417–35.

Houston, Lara, Jennifer Gabrys, and Helen Pritchard. "Breakdown in the Smart City: Exploring Workarounds with Urban-Sensing Practices and Technology." *Science, Technology, & Human Values* 44, no. 5 (2019): 843–70. https://doi.org/10.1177/01622 43919852677.

How to Work Together. http://howtoworktogether.org/.

Howarth, Robert W., Renee Santoro, and Anthony Ingraffea. "Methane and the Greenhouse-Gas Footprint of Natural Gas from Shale Formations." *Climatic Change* 106, no. 4 (2011): 679–90.

Howe, Cymene. "Sensing Asymmetries in Other-than-Human Forms." *Science, Technology, & Human Values* 44, no. 5 (2019): 900–910.

Hurdle, Jon. "PA Expands Particulate Monitoring as Federal Study Finds High Level in One Location." *State Impact: Pennsylvania (NPR)*, May 5, 2016. https://stateimpact .npr.org/pennsylvania/2016/05/05/pa-expands-particulate-monitoring-as-federal -study-finds-high-level-in-one-location.

Identifying Violations Affecting Neighborhoods. https://ivan-imperial.org/.

"India: Health Emergency Declared as Toxic Air Shrouds New Delhi." *Democracy Now*, November 8, 2017. https://www.democracynow.org/2017/11/8/headlines/india_health _emergency_declared_as_toxic_air_shrouds_new_delhi.

Ingraffea, Anthony R., Martin T. Wells, Renee L. Santoro, and Seth B. C. Shonkoff. "Assessment and Risk Analysis of Casing and Cement Impairment in Oil and Gas Wells in Pennsylvania, 2000–2012." *Proceedings of the National Academy of Sciences of the United States of America (PNAS)* 111, no. 30 (2014): 10955–60.

Institute of Technology in the Public Interest. https://github.com/InstitutefTiPI.

International Agency for Research on Cancer, World Health Organization. "IARC: Diesel Engine Exhaust Carcinogenic." Press Release no. 213. June 12, 2012. www.iarc.fr/en/ media-centre/pr/2012/pdfs/pr213_E.pdf.

Irani, Lilly. "Hackathons and the Making of Entrepreneurial Citizenship." *Science, Technology, & Human Values* 40, no. 5 (2015): 799–824.

Irwin, Alan. "Citizen Science and Scientific Citizenship: Same Words, Different Meanings?" In *Science Communication Today: Current Strategies and Means of Action*, edited by Berhand Schiele, Joëlle Le Marec, and Patrick Baranger, 29–38. Nancy: Nancy University, 2015.

Irwin, Alan. *Citizen Science: A Study of People, Expertise and Sustainable Development*. Abingdon, UK: Routledge, 1995.

Irwin, Alan. "Constructing the Scientific Citizen: Science and Democracy in the Biosciences." *Public Understanding of Science* 10, no. 1 (2001): 1–18.

Isin, Engin. *Being Political: Genealogies of Citizenship*. Minneapolis: University of Minnesota Press, 2002.

Isin, Engin, and Evelyn Ruppert. *Being Digital Citizens*. 2nd ed. London: Rowman and Littlefield, 2020.

James, William. *Essays in Radical Empiricism*. 1912. Lincoln: University of Nebraska Press, 1996.

James, William. "The Moral Philosopher and the Moral Life." In *The Will to Believe and Other Essays in Popular Philosophy*, 184–215. 1897. New York: Dover, 1956.

James, William. *A Pluralistic Universe*. 1909. Lincoln: University of Nebraska Press, 1996.

James, William. *Pragmatism and Other Writings*. New York: Penguin Books, 2000.

Jasanoff, Sheila. "Beyond Epistemology: Relativism and Engagement in the Politics of Science." *Social Studies of Science* 26, no. 2 (1996): 393–418.

Jasanoff, Sheila. "Technologies of Humility: Citizen Participation in Governing Science." *Minerva* 41 (2003): 223–44.

Jencks, Charles, and Nathan Silver. *Adhocism: The Case for Improvisation*. Expanded and updated ed. Cambridge, Mass.: MIT Press, 2013.

Kama, Kärg. "Resource-Making Controversies: Knowledge, Anticipatory Politics, and Economization of Unconventional Fossil Fuels." *Progress in Human Geography* 44, no. 2 (2020): 333–56.

Kaplan, Caren. "Atmospheric Politics: Protest Drones and the Ambiguity of Airspace." *Digital War* 1 (2020): 50–57.

Karban, Richard. *Plant Sensing and Communication*. Chicago: University of Chicago Press, 2015.

Karvinen, Kimmo, and Tero Karvinen. *Getting Started with Sensors*. Sebastopol, Calif.: Maker Media, 2014.

Kassotis, Christopher D., Donald E. Tillitt, J. Wade Davis, Annette M. Hormann, and Susan C. Nagel. "Estrogen and Androgen Receptor Activities of Hydraulic Fracturing Chemicals and Surface and Ground Water in a Drilling-Dense Region." *Endocrinology* 155, no. 3 (2014): 897–907.

Kay, Samuel, Bo Zhao, and Daniel Sui. "Can Social Media Clear the Air? A Case Study of the Air Pollution Problem in Chinese Cities." *The Professional Geographer* 67, no. 3 (2015): 351–63.

Kelty, Christopher, Aaron Panofsky, Morgan Currie, Roderic Crooks, Seth Erickson, Patricia Garcia, Michael Wartenbe, and Stacy Wood. "Seven Dimensions of Contemporary Participation Disentangled." *Journal of the Association for Information Science and Technology* 66, no. 3 (2015): 474–88.

Kenens, Joke, Michiel Van Oudheusden, Go Yoshizawa, and Ine Van Hoyweghen. "Science by, with and for Citizens: Rethinking 'Citizen Science' after the 2011 Fukushima Disaster." *Palgrave Communications* 6, no. 1 (2020): 1–8.

Kenner, Alison. *Breathtaking: Asthma Care in a Time of Climate Change*. Minneapolis: University of Minnesota Press, 2018.

Kessel, Jonah M., and Hiroko Tabuchi. "It's a Vast, Invisible Climate Menace: We Made It Visible." *New York Times*, December 12, 2019. https://www.nytimes.com/interactive/2019/12/12/climate/texas-methane-super-emitters.html

Kimmerer, Robin Wall. *Braiding Sweetgrass: Indigenous Wisdom, Scientific Knowledge, and the Teachings of Plants*. Minneapolis: Milkweed Editions, 2013.

Kimmerer, Robin Wall. "The Intelligence of Plants." *On Being with Krista Tippett*, February 25, 016, last updated August 20, 2020. https://onbeing.org/programs/robin-wall-kimmerer-the-intelligence-of-plants/.

Kimura, Aya H. "Citizen Science in Post-Fukushima Japan: The Gendered Scientization of Radiation Measurement." In "Conceptualizing Justice and Counter-Expertise," special issue of *Science as Culture* 28, no. 3 (2019): 327–50.

Kimura, Aya H., and Abby Kinchy. "Citizen Science: Probing the Virtues and Contexts of Participatory Research." *Engaging Science, Technology, and Society* 2 (2016): 331–61.

Kinchy, Abby, Kirk Jalbert, and Jessica Lyons. "What Is Volunteer Water Monitoring Good For? Fracking and the Plural Logics of Participatory Science." *Political Power and Social Theory* 27, no. 2 (2014): 259–89.

Kirk, Andrew G. *Counterculture Green: The "Whole Earth Catalog" and American Environmentalism*. Lawrence: University Press of Kansas, 2007.

Kleingeld, Pauline. *Kant and Cosmopolitanism: The Philosophical Ideal of World Citizenship*. Cambridge: Cambridge University Press, 2011.

Kohn, Eduardo. *How Forests Think: Toward an Anthropology beyond the Human*. Berkeley: University of California Press, 2013.

Kravets, David. "Law Making It Illegal to Collect Data, Photo of Open Land Hangs in Balance." *Ars Technica*, September 11, 2017. https://arstechnica.com/tech-policy/2017/09/ag-gag-law-gets-taken-to-the-slaughterhouse.

Kuchinskaya, Olga. "Citizen Science and the Politics of Environmental Data." *Science, Technology, & Human Values* 44, no. 5 (2019): 871–80.

Kukutai, Tahu, and John Taylor, eds. *Indigenous Data Sovereignty: Toward an Agenda*. Canberra: Australian National University Press, 2016.

LA Crypto Crew. "How to Become Anonymous Online." *Hyperallergic*, December 2, 2016. https://hyperallergic.com/342262/a-guide-to-becoming-anonymous-online.

LaDuke, Winona, and Deborah Cowen. "Beyond Wiindigo Infrastructure." *South Atlantic Quarterly* 119, no. 2 (2020): 243–68.

Lane, Stuart N., Nicholas Odoni, Catharina Landström, Sarah J. Whatmore, Neil Ward, and Susan Bradley. "Doing Flood Risk Science Differently: An Experiment in Radical Scientific Method." *Transactions of the Institute of British Geographers* 36 (2011): 15–36.

Lapoujade, David. *William James: Empiricism and Pragmatism*. Durham, N.C.: Duke University Press, 2019.

Larkin, Brian. "The Politics and Poetics of Infrastructure." *Annual Review of Anthropology* 42, no. 3 (2013): 327–43.

Latour, Bruno. "Atmosphère, Atmosphère." In *Olafur Eliasson: The Weather Project*, edited by Susan May, 29–41. London: Tate, 2014,

Latour, Bruno. "The Berlin Key, or How to Do Words with Things." In *Matter, Materiality, and Modern Culture*, edited by P. M. Graves-Brown, 10–21. London: Routledge, 1991.

Latour, Bruno. *Facing Gaia: Eight Lectures on the New Climatic Regime*. Cambridge, UK: Polity, 2017.

Latour, Bruno. *Science in Action: How to Follow Scientists and Engineers through Society*. Cambridge, Mass.: Harvard University Press, 1987.

Latour, Bruno. "Tarde's Idea of Quantification." In *The Social after Gabriel Tarde: Debates and Assessments*, edited by Matei Candea, 145–62. London: Routledge, 2009.

Lave, Rebecca. "The Future of Environmental Expertise." *Annals of the Association of American Geographers* 105, no. 2 (2015): 244–52.

Lave, Rebecca, and Brian Lutz. "Hydraulic Fracturing: A Critical Physical Geography Review." *Geography Compass* 8, no. 10 (2014): 739–54.

Laville, Sandra. "Air Pollution a Cause in Girl's Death, Coroner Rules in Landmark Case." *The Guardian*, December 16, 2020. https://www.theguardian.com/environment/20 20/dec/16/girls-death-contributed-to-by-air-pollution-coroner-rules-in-landmark-case.

Laville, Sandra. "UK Waste Incinerators Three Times as Likely to Be in Deprived Areas." *The Guardian*, July 31, 2020. https://www.theguardian.com/environment/2020/ jul/31/uk-waste-incinerators-three-times-more-likely-to-be-in-deprived-areas.

Law, John. "What's Wrong with a One-World World?" *Distinktion: Scandinavian Journal of Social Theory* 16, no. 1 (2015): 126–39.

Law, John, and Solveig Joks. "Indigeneity, Science, and Difference: Notes on the Politics of How." *Science, Technology, & Human Values* 44, no. 3 (2018): 424–47.

Lefebvre, Henri. *Writings on Cities: Henri Lefebvre*. Selected, translated, and introduced by Eleonore Kofman and Elizabeth Lebas. Oxford: Blackwell, 1996.

Lelieveld, Jos, Klaus Klingmüller, Andrea Pozzer, Ulrich Pöschl, Mohammed Fnais, Andreas Daiber, and Thomas Münzel. "Cardiovascular Disease Burden from Ambient Air Pollution in Europe Reassessed Using Novel Hazard Ratio Functions." *European Heart Journal* 40, no. 20 (2019): 1590–96.

Lewisham Council. "Tidemill Site Development—Questions and Answers." https://lew isham.gov.uk/tidemill.

Lewisham Council. "What We Are Doing to Improve Air Quality in Lewisham." https:// lewisham.gov.uk/myservices/environment/air-pollution/what-we-are-doing-to-im prove-air-quality-in-lewisham.

Lezaun, Javier, Noortje Marres, and Manuel Tironi. "Experiments in Participation." In *The Handbook of Science and Technology Studies*, edited by Ulrike Felt, Rayvon Fouché, Clark A. Miller, and Laurel Smith-Doerr, 195–222. Cambridge, Mass.: MIT Press, 2016.

Li, Xiaoyue, and Bryan Tilt. "Public Engagements with Smog in Urban China: Knowledge, Trust, and Action." *Environmental Science & Policy* 92 (2019): 220–27.

Lippert, Ingmar, and Helen Verran. "After Numbers? Innovations in Science and Technology Studies' Analytics of Numbers and Numbering." *Science & Technology Studies* 31, no. 4 (2018): 2–12.

Llewellyn, Garth T., Frank Dorman, J. L. Westland, D. Yoxtheimer, Paul Grieve, Todd Sowers, E. Humston-Fulmer, and Susan L. Brantley. "Evaluating a Groundwater Supply Contamination Incident Attributed to Marcellus Shale Gas Development." *Proceedings of the National Academy of Sciences of the United States of America (PNAS)* 112, no. 20 (2015): 6325–30.

London Air. "Lewisham." https://www.londonair.org.uk/london/asp/publicbulletin.asp? la_id=23&MapType=Google.

London Inner South Coroner's Court. "Inquest Touching the Death of Ella Roberta Adoo Kissi-Debrah." https://www.innersouthlondoncoroner.org.uk/news/2020/nov/inquest -touching-the-death-of-ella-roberta-adoo-kissi-debrah.

Loukissas, Yanni Alexander. *All Data Are Local: Thinking Critically in a Data-Driven Society*. Cambridge, Mass.: MIT Press, 2019.

Lowe, Lisa. *The Intimacies of Four Continents*. Durham, N.C.: Duke University Press, 2015.

Loxham, Matthew, Donna E. Davies, and Stephen T. Holgate. "The Health Effects of Fine Particulate Air Pollution: The Harder We Look, the More We Find." *BMJ* 367, no. 16609 (2019): 1–12.

Lupton, Deborah. *The Quantified Self: A Sociology of Self-Tracking.* Cambridge, UK: Polity, 2016.

Lynch, M., and S. Cole. "Science and Technology Studies on Trial: Dilemmas of Expertise." *Social Studies of Science* 35, no. 2 (2005): 269–311.

Lyons, Kristina M. *Vital Decomposition: Soil Practitioners and Life Politics.* Durham, N.C.: Duke University Press, 2020.

Macey, Gregg P., Ruth Breech, Mark Chernaik, Caroline Cox, Denny Larson, Deb Thomas, and David O. Carpenter. "Air Concentrations of Volatile Compounds near Oil and Gas Production: A Community-Based Exploratory Study." *Environmental Health* 13, no. 82 (2014): 1–18. http://www.ehjournal.net/content/13/1/82.

Mackenzie, Adrian. *Wirelessness: Radical Empiricism in Network Cultures.* Cambridge, Mass.: MIT Press, 2010.

Maguire, James, and Britt Ross Winthereik. "Digitalizing the State: Data Centres and the Power of Exchange." *Ethnos* 5 (2019). https://doi.org/10.1080/00141844.2019.1660391.

Majaca, Antonia, and Luciana Parisi. "The Incomputable and Instrumental Possibility." *E-Flux Journal* 77 (November 2016). https://www.e-flux.com/journal/77/76322/the-incomputable-and-instrumental-possibility/.

Makea Tu Vida. http://www.makeatuvida.net/.

Mapping for Change. "Science in the City: Barbican Citizen Science Project." March 11, 2015. https://www.youtube.com/watch?v=YvSuPCxli88.

Mapping for Change. "Science in the City: Barbican Report." 2015. https://www.barbicanassociation.co.uk/wp-content/uploads/2015/02/airquality1.pdf.

Marcellus Gas. https://www.marcellusgas.org.

Mayor of London. London Data Store. "2011 Census Ethnic Group Fact Sheets." https://data.london.gov.uk/dataset/2011-census-ethnic-group-fact-sheets.

Mbembe, Achille. "The Universal Right to Breathe." Translated by Carolyn Shread. *Critical Inquiry,* "In the Moment," April 13, 2020. https://critinq.wordpress.com/2020/04/13/the-universal-right-to-breathe/.

McClintock, Nathan. "Radical, Reformist, and Garden-Variety Neoliberal: Coming to Terms with Urban Agriculture's Contradictions." *Local Environment* 19, no. 2 (2014): 147–71.

McCormack, Derek P. *Atmospheric Things: On the Allure of Elemental Envelopment.* Durham, N.C.: Duke University Press, 2018.

McGlotten, Shaka. "Black Data." In *No Tea, No Shade: New Writings in Black Queer Studies,* edited by E. Patrick Johnson, 262–86. Durham, N.C.: Duke University Press, 2016.

McKittrick, Katherine. *Dear Science and Other Stories.* Durham, N.C.: Duke University Press, 2021.

McKittrick, Katherine, ed. *Sylvia Wynter: On Being Human as Praxis.* Durham, N.C.: Duke University Press, 2015.

Meng, Amanda, and Carl DiSalvo. "Grassroots Resource Mobilization through Counter-Data Action." *Big Data & Society* 5, no. 2 (2018). https://doi.org/10.1177%2F2053951718796862.

Mertia, Sandeep, ed. *Lives of Data: Essay on Computational Cultures from India.* Amsterdam: Institute of Network Cultures, 2020.

Michael, Mike. "Publics Performing Publics: Of PiGs, PiPs and Politics." *Public Understanding of Science* 18, no. 5 (2009): 617–31.

Mignolo, Walter. "On Pluriversality." October 20, 2013. http://waltermignolo.com/on
-pluriversality.

Milan, Stefania, and Emiliano Treré. "Big Data from the South(s): Beyond Data Univer-
salism." *Television & New Media* 20, no. 4 (2019): 319–35.

Mitchell, Gordon, and Danny Dorling. "An Environmental Justice Analysis of British Air
Quality." *Environment and Planning A* 35, no. 5 (2003): 909–29.

Mody, Cyrus. *Instrumental Community: Probe Microscopy and the Path to Nanotechnology.*
Cambridge, Mass.: MIT Press, 2011.

Mol, Annemarie. *The Body Multiple: Ontology in Medical Practice.* Durham, N.C.: Duke
University Press, 2002.

Mol, Annemarie. *The Logic of Care: Health and the Problem of Patient Choice.* London:
Routledge, 2008.

Mol, Annemarie, Ingunn Moser, and Jeanette Pols. "Care: Putting Practice into Theory."
In *Care in Practice: On Tinkering in Clinics, Homes and Farms,* edited by Annemarie
Mol, Ingunn Moser, and Jeanette Pols, 7–26. Bielefeld, Germany: Transcript, 2010.

Moore, Chris. "Air Impacts of Gas Shale Extraction and Distribution." *Workshop on Risks
of Unconventional Shale Gas Development.* May 30–31, 2013. http://sites.nationalacad
emies.org/cs/groups/dbassesite/documents/webpage/dbasse_083402.pdf.

Morello-Frosch, Rachel, Manuel Pastor, and James Sadd. "Environmental Justice and
Southern California's 'Riskscape': The Distribution of Air Toxics Exposures and Health
Risks among Diverse Communities." *Urban Affairs Review* 36, no. 4 (2001): 551–78.

Morgan, Ben. "'It's Time to Act Now,' Say Researchers Who Found Pollution at Six Times
World Safety Limit in Deptford and New Cross." *Evening Standard,* November 14,
2017. https://www.standard.co.uk/news/london/it-s-time-to-act-now-say-researchers
-who-found-pollution-at-six-times-world-safety-limit-in-deptford-and-new-cross-a36
90451.html.

Müller, Jan-Werner. "Behind the New German Right." *New York Review of Books,* April 14,
2016. http://www.nybooks.com/daily/2016/04/14/behind-new-german-right-afd.

Murphy, Michelle. *Sick Building Syndrome and the Problem of Uncertainty.* Durham, N.C.:
Duke University Press, 2006.

Museum of London. "How to Grow Your Own Air Quality Garden." April 23, 2018.
https://www.museumoflondon.org.uk/discover/phyto-sensor-toolkit-citizen-sense
-air-pollution.

Nadim, Tahani. "Blind Regards: Troubling Data and Their Sentinels." *Big Data & Society*
3, no. 2 (2016): 1–6.

National Resources Defense Council. "NRDC Policy Basics: Fracking." 2013. http://www
.nrdc.org/legislation/policy-basics/files/policy-basics-fracking-FS.pdf.

Nelson, Alondra. *Body and Soul: The Black Panther Party and the Fight against Medical
Discrimination.* Minneapolis: University of Minnesota Press, 2011.

Neville, Kate J., Jennifer Baka, Shanti Gamper-Rabindran, Karen Bakker, Stefan Andreas-
son, Avner Vengosh, Alvin Lin, Jewellord Nem Singh, and Erika Weinthal. "Debating
Unconventional Energy: Social, Political, and Economic Implications." *Annual Review
of Environment and Resources* 42 (2017): 241–66.

New York Times. "A Methane Leak, Seen from Space, Proves to Be Far Larger than
Thought." December 16, 2019. https://www.nytimes.com/2019/12/16/climate/meth
ane-leak-satellite.html.

Nimis, Pier Luigi, Christoph Scheidegger, and Patricia Wolseley, eds. *Monitoring with Lichens—Monitoring Lichens.* Amsterdam: Kluwer Academic Publishers, 2002.

Noble, Safiya Umoja. *Algorithms of Oppression: How Search Engines Reinforce Racism.* New York: New York University Press, 2018.

Nold, Christian. "Device Studies of Participatory Sensing: Ontological Politics and Design Interventions." PhD diss., University College London, 2017.

Noor, Poppy. "Housing Approved Despite Pollution Warning to Keep Windows Shut." *The Guardian,* April 12, 2019. https://www.theguardian.com/environment/2019/apr/12/london-housing-approved-in-area-with-illegal-pollution-levels-lewisham.

Olsen, Erik. "Natural Gas and Polluted Air." *New York Times Video,* February 2, 2011. https://www.nytimes.com/video/us/100000000650773/natgas.html.

Onuoha, Mimi, and Mother Cyborg [Diana Nucera]. "A People's Guide to AI." Allied Media, 2018. https://alliedmedia.org/resources/peoples-guide-to-ai.

Open Policy Making Toolkit. https://www.gov.uk/guidance/open-policy-making-toolkit.

OSBeehives. "BuzzBox." https://www.osbeehives.com/.

Osborn, Stephen G., Avner Vengosh, Nathaniel R. Warner, and Robert B. Jackson. "Methane Contamination of Drinking Water Accompanying Gas-Well Drilling and Hydraulic Fracturing." *Proceedings of the National Academy of Sciences (PNAS)* 108, no. 2 (2011): 8172–76.

Ottinger, Gwen. "Buckets of Resistance: Standards and the Effectiveness of Citizen Science." *Science, Technology, & Human Values* 35, no. 2 (2010): 244–70.

Our Data Bodies Project. https://www.odbproject.org/our-data-bodies-project/.

PA Media. "Inquest to Determine If London Air Pollution Caused Child's Death." *The Guardian,* December 19, 2019. https://www.theguardian.com/environment/2019/dec/17/inquest-air-pollution-ella-kissi-debrah-death.

Parasie, Sylvain, and François Dedieu. "What Is the Credibility of Citizen Data Based on?" *Revue d'Anthropologie des Connaissances* 13, no. 4 (2019): 1035–62.

Parau, Cristina Elena, and Jerry Wittmeier Bains. "Europeanisation as Empowerment of Civil Society: All Smoke and Mirrors?" In *Civil Society and Governance in Europe: From National to International Linkages,* edited by William A. Maloney and Jan W. van Deth, 109–26. Cheltenham: Edward Elgar Publishing, 2008.

Peabody. "Frankham Street Development." https://www.peabody.org.uk/homes-in-development/lewisham/frankham-street-development.

Pearce, Jamie, Simon Kingham, and Peyman Zawar-Reza. "Every Breath You Take? Environmental Justice and Air Pollution in Christchurch, New Zealand." *Environment and Planning A* 38, no. 5 (2006): 919–38.

Peirce, Charles Sanders. "The Fixation of Belief." In *Chance, Love, and Logic: Philosophical Essays,* 7–31. 1877. Lincoln: University of Nebraska Press, 1998.

Pennsylvania Alliance for Clean Water and Air. "List of the Harmed." http://pennsylvaniaallianceforcleanwaterandair.wordpress.com/the-list.

Pennsylvania Department of Environmental Protection. "DEP Expands Particulate Matter Air Monitoring Network." Press Release. April 27, 2016. http://www.media.pa.gov/Pages/DEP_details.aspx?newsid=629.

Perng, Sung-Yueh, and Sophia Maalsen. "Civic Infrastructure and the Appropriation of the Corporate Smart City." *Annals of the American Association of Geographers* 110, no. 2 (2020): 507–15.

Petryna, Adriana. *Life Exposed: Biological Citizens after Chernobyl.* 2nd ed. Princeton, N.J.: Princeton University Press, 2013.

Philip, Kavita, Lily Irani, and Paul Dourish. "Postcolonial Computing: A Tactical Survey." *Science, Technology, & Human Values* 37, no. 1 (2012): 3–29.

Phillips, Susan. "Burning Question: What Would Life Be Like without the Halliburton Loophole?" *State Impact: Pennsylvania (NPR)*, December 5, 2011. https://stateimpact .npr.org/pennsylvania/2011/12/05/burning-question-what-would-life-be-like-without -the-halliburton-loophole.

Pickering, Andrew. *The Mangle of Practice: Time, Agency, and Science.* Chicago: University of Chicago Press, 1995.

Pidot, Justin. "Forbidden Data: Wyoming Just Criminalized Citizen Science." *Slate*, May 11, 2015. http://www.slate.com/articles/health_and_science/science/2015/05/wyoming _law_against_data_collection_protecting_ranchers_by_ignoring_the.html.

Plantin, Jean-Christophe. "The Politics of Mapping Platforms: Participatory Radiation Mapping after the Fukushima Daiichi Disaster." *Media, Culture & Society* 37, no. 6 (2015): 904–21.

"Plants Employed as Sensing Devices." https://pleased-fp7.eu/.

Plume Labs. "Clean Air, Together—Flow by Plume Labs." September 26, 2017. https:// www.youtube.com/watch?v=uJtahMBOn6g.

Plumer, Brad, and Nadja Popovich. "How Decades of Racist Housing Policy Left Neighborhoods Sweltering." *New York Times*, August 24, 2020. https://www.nytimes.com/ interactive/2020/08/24/climate/racism-redlining-cities-global-warming.html.

Pollock, Anne, and Banu Subramaniam. "Resisting Power, Retooling Justice: Promises of Feminist Postcolonial Technosciences." *Science, Technology, & Human Values* 41, no. 6 (2016): 951–66.

Povinelli, Elizabeth. "The Toxic Earth and the Collapse of Political Concepts." Keynote lecture at "Critical Ecologies." Goldsmiths, University of London, March 17, 2018.

Powell, Alison. *Undoing Optimization: Civic Action in Smart Cities.* New Haven, Conn.: Yale University Press, 2021.

Pritchard, Helen. "The Animal Hacker." PhD diss., London: Queen Mary University of London, 2018.

Pritchard, Helen, and Jennifer Gabrys, "From Citizen Sensing to Collective Monitoring: Working through the Perceptive and Affective Problematics of Environmental Pollution." *Geohumanities* 2, no. 2 (2016): 354–71.

Pritchard, Helen, Jennifer Gabrys, and Lara Houston. "Re-calibrating DIY: Testing Digital Participation across Dust Sensors, Fry Pans, and Environmental Pollution." *New Media and Society* 20, no. 12 (2018): 4533–52. https://doi.org/10.1177/1461444818777473.

Puig de la Bellacasa, Maria. "Making Time for Soil: Technoscientific Futurity and the Pace of Care." *Social Studies of Science* 45, no. 5 (2015): 691–716.

Puig de la Bellacasa, Maria. "Matters of Care in Technoscience: Assembling Neglected Things." *Social Studies of Science* 41, no. 1 (2011): 85–106.

Pulido, Laura. "Flint, Environmental Racism, and Racial Capitalism." *Capitalism Nature Socialism* 27, no. 3 (2016): 1–16.

Radiation Watch. "Pocket Geiger." http://www.radiation-watch.org/; https://www.spark fun.com/products/14209.

Rankine, Claudia. *Citizen: An American Lyric.* Minneapolis: Graywolf Press, 2014.

Ratto, Matthew, and Megan Boler, eds. *DIY Citizenship: Critical Making and Social Media.* Cambridge, Mass.: MIT Press, 2014.

rawrdong. "How to Build a Portable, Accurate, Low Cost, Open Source Air Particle Counter." http://www.instructables.com/id/How-to-Build-a-Portable-Accurate-Low-Cost-Open-Sou.

The Real News Network. "Police Are Shooting Down Aerial Drones over Standing Rock Demonstration." October 30, 2016. https://www.youtube.com/watch?v=guUQ5zEjP9k.

Reardon, Jenny, Jacob Metcalf, Martha Kenney, and Karen Barad. "Science and Justice: The Trouble and the Promise." *Catalyst: Feminism, Theory, Technoscience* 1, no. 1 (2015): 1–49.

Renzi, Alessandra, and Ganaele Langlois. "Data Activism." In *Compromised Data: New Paradigms in Social Media Theory and Methods,* edited by Greg Elmer, Ganaele Langlois, and Joanna Redden, 202–25. London: Bloomsbury, 2015.

Rosner, Daniela K. *Critical Fabulations: Reworking the Methods and Margins of Design.* Cambridge, Mass.: MIT Press, 2018.

Roter, Rebecca. "Breathe Easy Susquehanna County." June 6, 2013. https://en-gb.facebook.com/BreatheEasySusq/posts/1657969169349779.

Royal College of Physicians. "Every Breath We Take: The Lifelong Impact of Air Pollution." *Report of a Working Party.* London: RCP, 2016. https://www.rcplondon.ac.uk/projects/outputs/every-breath-we-take-lifelong-impactair-pollution.

Ruser, Nathan. "How to Scrape Interactive Geospatial Data." September 5, 2018. https://www.bellingcat.com/resources/how-tos/2018/09/05/scrape-interactive-geospatial-data.

Rusert, Britt. *Fugitive Science: Empiricism and Freedom in Early African American Culture.* New York: New York University Press, 2017.

Save Reginald! Save Tidemill! "Destruction of Deptford's Much Loved Community Garden Ordered by Peabody and Sanctioned by Lewisham Council." February 27, 2019. https://www.facebook.com/savetidemill/videos/613634585727416/.

Save Reginald! Save Tidemill! "Help Us Save Reginald House and Tidemill Wildlife Garden." https://www.crowdjustice.com/case/save-reginald-save-tidemill/.

Schrader, Astrid. "Responding to *Pfiesteria piscicida* (the Fish Killer): Phantomatic Ontologies, Indeterminacy and Responsibility in Toxic Microbiology." *Social Studies of Science* 40, no. 2 (2010): 275–306.

Schuppli, Susan. *Material Witness: Media, Forensics, Evidence.* Cambridge, Mass.: MIT Press, 2020.

Scroggins, Vera. https://www.youtube.com/channel/UCfbfGPFJn5t3HvJcRNTvJxQ.

Seeed Studio. "Grove Smart Plant Care Kit for Arduino." https://www.seeedstudio.com/Grove-Smart-Plant-Care-Kit-for-Arduino-p-2528.html.

Seitz, David K. Interview with Lauren Berlant. "On Citizenship and Optimism." *Society + Space,* March 22, 2013. http://societyandspace.org/2013/03/22/on-citizenship-and-optimism.

Shapin, Steven, and Simon Schaffer. *Leviathan and the Air-Pump: Hobbes, Boyle, and the Experimental Life.* Princeton, N.J.: Princeton University Press, 1985.

Sharpe, Christina. *In the Wake: On Blackness and Being.* Durham, N.C.: Duke University Press, 2016.

Sharpe, Christina. "The Weather." *The New Inquiry*, January 19, 2017. https://thenewin quiry.com/the-weather/.

Shaviro, Steven. *The Universe of Things: On Speculative Realism*. Minneapolis: University of Minnesota Press, 2014.

Shaviro, Steven. *Without Criteria: Kant, Whitehead, Deleuze, and Aesthetics*. Cambridge, Mass.: MIT Press, 2009.

Sheikh, Shela. "The Future of the Witness: Nature, Race, and More-than-Human Environmental Publics." *Kronos* 44, no. 1 (2018): 145–62.

Shelton, Taylor, and Thomas Lodato. "Actually Existing Smart Citizens: Expertise and (Non)participation in the Making of the Smart City." *City* 23, no. 1 (2019): 35–52.

Shinyei. Particle Sensor Unit PPD42NJ. https://www.shinyei.co.jp/stc/eng/products/ optical/ppd42nj.html.

Siisip Geniusz, Mary. *Plants Have So Much to Give Us, All We Have to Do Is Ask: Anishinaabe Botanical Teachings*. Minneapolis: University of Minnesota Press, 2015.

Simmons, Kristen. "Settler Atmospherics." *Cultural Anthropology*. Member Voices: Fieldsights, November 20, 2017. https://culanth.org/fieldsights/settler-atmospherics.

Simondon, Gilbert. *L'individuation à la lumière des notions de forme et d'information*. Grenoble: Collection Krisis, Éditions Jérôme Millon, 2005.

Simondon, Gilbert. *Individuation in Light of Notions of Form and Information*. Translated by Taylor Adkins. Minneapolis: University of Minnesota Press, 2020.

Simondon, Gilbert. *On the Mode of Existence of Technical Objects*. Translated by Cécile Malaspina and John Rogove. 1958. Minneapolis: Univocal, 2017.

Simpson, Audra. *Mohawk Interruptus: Political Life across the Borders of Settler States*. Durham, N.C.: Duke University Press, 2014.

Simpson, Leanne Betasamosake. *As We Have Always Done: Indigenous Freedom through Radical Resistance*. Minneapolis: University of Minnesota Press, 2017.

Simpson, Leanne Betasamosake. "Land as Pedagogy: Nishnaabeg Intelligence and Rebellious Transformation." *Decolonization: Indigeneity, Education & Society* 3, no. 3 (2014): 1–25.

Singh, Julietta. *Unthinking Mastery: Dehumanism and Decolonial Entanglements*. Durham, N.C.: Duke University Press, 2018.

Sloterdijk, Peter. *Terror from the Air*. Translated by Amy Patton and Steve Corcoran. Cambridge, Mass.: MIT Press, 2009.

Smith, Linda Tuhiwai. *Decolonizing Methodologies: Research and Indigenous Peoples*. 2nd ed. London: Zed Books, 2012.

Smith, Mistinguette. "Wildness." Center for Humans and Nature. March 20, 2017. https://www.youtube.com/watch?v=p3A044lnFIU.

Soeder, Daniel J., and William M. Kappel. "Water Resources and Natural Gas Production from the Marcellus Shale." Fact Sheet 2009-3032 (May 2009). US Geological Survey. https://pubs.usgs.gov/fs/2009/3032/pdf/FS2009-3032.pdf.

Soon, Winnie, and Geoff Cox. *Aesthetic Programming: A Handbook of Software Studies*. Open Humanities Press, 2020.

South East London Combined Heat and Power. "History." https://www.selchp.com/about -selchp/history/.

Southwest Pennsylvania Environmental Health Project. "Citizen Science Toolkit." 2017. http://www.environmentalhealthproject.org/citizen-science-toolkit.

Spencer, Michaela, Endre Dányi, and Yasunori Hayashi. "Asymmetries and Climate Futures: Working with Waters in an Indigenous Australian Settlement." *Science, Technology, & Human Values* 44, no. 5 (2019): 786–813.

Spice, Byron, and Suzanne Thinnes. "CMU, Airviz Will Make Air Quality Monitors Available at Public Libraries Nationwide." *Carnegie Mellon News*, March 15, 2016. https://www.cmu.edu/news/stories/archives/2016/march/air-monitors-in-libraries.html.

Spivak, Gayatri Chakravorty. *Imperatives to Re-imagine the Planet/Imperative zur Neuerfindung des Planeten*. Vienna: Passagen, 1999.

sspence. "Earthquake/Vibration Sensor." Instructables. April 28, 2016. http://www.instructables.com/id/Earthquake-Vibration-Sensor.

State Impact. "The Marcellus Shale, Explained." http://stateimpact.npr.org/pennsylvania/tag/marcellus-shale.

Steele, Jess, ed. *Deptford Creek Surviving Regeneration*. London: Deptford Forum Publishing, 1998.

Steinzor, Nadia. "Community Air Monitoring of Oil and Gas Pollution: A Survey of Issues and Technologies." *Earthworks*, February 29, 2016. https://www.earthworksaction.org/publications/community_air_monitoring_of_oil_and_gas_pollution_a_survey_of_issues_and_te/.

Stengers, Isabelle. *Cosmopolitics I*. Translated by Robert Bononno. 1997. Minneapolis: University of Minnesota Press, 2010.

Stengers, Isabelle. *Cosmopolitics II*. Translated by Robert Bononno. 1997. Minneapolis: University of Minnesota Press, 2011.

Stengers, Isabelle. "Including Nonhumans in Political Theory: Opening Pandora's Box?" In *Political Matter: Technoscience, Democracy, and Public Life*, edited by Bruce Braun and Sarah Whatmore, 3–34. Minneapolis: University of Minnesota Press, 2010.

Stengers, Isabelle. *Thinking with Whitehead: A Free and Wild Creation of Concepts*. Translated by Michael Chase. 2002. Cambridge, Mass.: Harvard University Press, 2011.

Stewart, Kathleen. "Atmospheric Attunements." *Environment and Planning D: Society and Space* 29 (2011): 445–53.

Strasser, Anita. "Tidemill Garden Is Part of the Cohesiveness of Deptford." October 22, 2018. https://deptfordischanging.wordpress.com/2018/10/.

Suchman, Lucy. *Human–Machine Reconfigurations: Plans and Situated Actions*. Cambridge: Cambridge University Press, 2007.

Suchman, Lucy, Randall Trigg, and Jeanette Blomberg. "Working Artefacts: Ethnomethods of the Prototype." *British Journal of Sociology* 53, no. 2 (2002): 163–79.

Swart, Dirk. "Egg Version One End of Life." May 19, 2017. https://shop.wickeddevice.com/2017/05/19/egg-version-one-end-of-life.

Sze, Julie. *Noxious New York: The Racial Politics of Urban Health and Environmental Justice*. Cambridge, Mass.: MIT Press, 2006.

Szeman, Imre, and Dominic Boyer, eds. *Energy Humanities: An Anthology*. Baltimore: Johns Hopkins University Press, 2017.

Tactical Tech. "Data Detox Kit." Produced for the Glass Room, London, November 2017. https://datadetox.myshadow.org/en/detox.

TallBear, Kim. "Why Interspecies Thinking Needs Indigenous Standpoints." Theorizing the Contemporary, *Cultural Anthropology*, April 24, 2011. https://culanth.org/fieldsights/260-why-interspecies-thinking-needs-indigenous-standpoints.

Tarantola, Andrew. "How to Erase Yourself from the Internet." *Gizmodo*, November 2, 2013. https://gizmodo.com/how-to-erase-yourself-from-the-internet-1456270634.

Taub, Liba. "Introduction: Reengaging with Instruments." *Isis* 102, no. 4 (2011): 689–96.

Taylor, Keenaga-Yamahtta. *How We Get Free: Black Feminism and the Combahee River Collective*. Chicago: Haymarket Books, 2017.

Taylor, Linnet. "What Is Data Justice? The Case for Connecting Digital Rights and Freedoms Globally." *Big Data & Society* 4, no. 2 (2017): 1–14.

Thames Tideway Tunnel. https://www.tideway.london/the-tunnel/.

Tobin, Olivia. "Extinction Rebellion Lewisham: London Roads Blocked Off by Climate Change Activists in Major Rush-Hour Protest." *Evening Standard*, June 14, 2019. https://www.standard.co.uk/news/london/extinction-rebellion-lewisham-protest-cli mate-change-activists-block-off-major-roads-in-south-london-a4167101.html.

Trafton, Anne. "Nanobionic Spinach Plants Can Detect Explosives." *MIT News*, October 31, 2016. https://news.mit.edu/2016/nanobionic-spinach-plants-detect-explosives-1031.

Tsing, Anna Lowenhaupt. *The Mushroom at the End of the World: On the Possibility of Life in Capitalist Ruins*. Princeton, N.J.: Princeton University Press, 2015.

Tuck, Eve, Mistinguette Smith, Allison M. Guess, Tavia Benjamin, and Brian K. Jones. "Geotheorizing Black/Land: Contestations and Contingent Collaborations." *Departures in Critical Qualitative Research* 3, no. 1 (2014): 52–74.

Turner, Fred. *From Counterculture to Cyberculture: Stewart Brand, the Whole Earth Network, and the Rise of Digital Utopianism*. Chicago: University of Chicago Press, 2006.

United Nations Economic Commission for Europe. "Public Participation." https://www .unece.org/ro/env/pp/welcome.html.

United Nations Human Rights Council. "Human Rights and the Environment." A/HRC/ RES/37/8. April 9, 2018. https://ap.ohchr.org/documents/dpage_e.aspx?si=A%2FHR C%2FRES%2F37%2F8.

United States Court of Appeals, Tenth Circuit. People for the Ethical Treatment of Animals Inc. Center for Food Safety v. Center for Agriculture and Food Systems First Amendment Legal Scholars. United States Court of Appeals, Tenth Circuit, No. 16– 8083 (Decided: September 7, 2017). http://caselaw.findlaw.com/us-10th-circuit/18732 99.html.

United States Energy Policy Act, 2005. https://www.gpo.gov/fdsys/pkg/PLAW-109publ58/ html/PLAW-109publ58.htm.

United States Environmental Protection Agency. "Air Sensor Toolbox." https://www.epa .gov/air-sensor-toolbox.

United States Environmental Protection Agency. "Care for Your Air: A Guide to Indoor Air Quality." https://www.epa.gov/sites/production/files/2014-08/documents/carefor yourair.pdf.

United States Environmental Protection Agency. "Draft Roadmap for Next Generation Air Monitoring." March 8, 2013. http://www.eunetair.it/cost/newsroom/03-US-EPA _Roadmap_NGAM-March2013.pdf.

van Haluwyn, Chantal, and C. M. van Herk. "Bioindication: The Community Approach." In *Monitoring with Lichens—Monitoring Lichens*, edited by Pier Luigi Nimis, Christoph Scheidegger, and Patricia A. Wolseley, 39–64. Amsterdam: Kluwer Academic Publishers, 2002.

Veith, Dave. "AWS IoT and Beehives." https://www.hackster.io/bees/aws-iot-and-beehives
-c59fff.

Verran, Helen. "The Changing Lives of Measures and Values: From Centre Stage in the
Fading 'Disciplinary' Society to Pervasive Background Instrument in the Emergent
'Control' Society." *Sociological Review* 59, issue: 2_suppl (2011): 60–72.

Vickers, Harriet. "The Battle for Deptford: Teaser." May 28, 2019. https://www.facebook
.com/battlefordeptford/videos/316074959305965/UzpfSTM2OTE4NDQ1Njg5OTc
1Mj02MDMzODk5NzY4MTI1MzE/.

Viveiros de Castro, Eduardo. *Cannibal Metaphysics: For a Post-structural Anthropology.*
Translated and edited by Peter Skafish. 2009. Minneapolis: Univocal, 2014.

Walker, Dawn, Eric Nost, Aaron Lemelin, Rebecca Lave, and Lindsey Dillon. "Practicing
Environmental Data Justice: From DataRescue to Data Together." *Geo: Geography and
Environment* 5, no. 2 (2018): 1–14.

Walker, Gordon. "Beyond Distribution and Proximity: Exploring the Multiple Spatialities
of Environmental Justice." *Antipode* 41, no. 4 (2009): 614–36.

Walker, Gordon, Douglas Booker, and Paul J. Young. "Breathing in the Polyrhythmic City:
A Spatiotemporal, Rhythmanalytic Account of Urban Air Pollution and Its Inequalities." *Environment and Planning C: Politics and Space* (2020). https://doi.org/10.1177%
2F2399654420948871.

Walton, Heather, David Dajnak, Sean Beevers, Martin Williams, Paul Watkiss, and
Alistair Hunt. "Understanding the Health Impacts of Air Pollution in London." King's
College London, produced for Transport for London and the Greater London Authority. July 14, 2015. https://www.london.gov.uk/sites/default/files/hiainlondon_kings
report_14072015_final.pdf.

Waterton, Claire, and Judith Tsouvalis. "'An Experiment with Intensities': Village Hall
Reconfigurings of the World within a New Participatory Collective." In *Remaking
Participation: Science, Environment, and Emergent Publics,* edited by Jason Chilvers and
Matthew Kearnes, 201–17. Abingdon, UK: Routledge, 2015.

Wenzel, Jennifer. "Reading Fanon Reading Nature." In *What Postcolonial Theory Doesn't
Say,* edited by Anna Bernard, Ziad Elmarsafy, and Stuart Murray, 185–201. Abingdon,
UK: Routledge, 2015.

West, Cornel. *The American Evasion of Philosophy: A Genealogy of Pragmatism.* Madison:
University of Wisconsin Press, 1989.

West Oakland Environmental Indicators Project. https://woeip.org.

"What You'll Need to Escape New York." *New York Times,* January 25, 2013. https://archive
.nytimes.com/www.nytimes.com/interactive/2013/01/27/nyregion/preppers-bug-out
-bag.html.

Whitehead, Alfred North. *Modes of Thought.* 1938. New York: The Free Press, 1966.

Whitehead, Alfred North. *Process and Reality.* 1929. New York: Free Press, 1985.

Whitehead, Alfred North. *Science and the Modern World.* New York: Free Press, 1925.

Whitehead, Mark. *State Science and the Skies: Governmentalities of the British Atmosphere.*
Malden, Mass.: Wiley-Blackwell, 2009.

Whyte, Kyle Powys. "The Dakota Access Pipeline, Environmental Injustice, and U.S.
Colonialism." *Red Ink: An International Journal of Indigenous Literature, Arts, & Humanities* 19, no. 1 (Spring 2017): 154–69.

Whyte, Kyle Powys. "Indigenous Science (Fiction) for the Anthropocene: Ancestral Dystopias and Fantasies of Climate Change Crises." *Environment and Planning E: Nature and Space* 1, nos. 1–2 (2018): 224–42.

Whyte, Kyle Powys. "Indigenous Women, Climate Change Impacts, and Collective Action." *Hypatia* 29, no. 3 (2014): 599–616.

Wicked Device. "Air Quality Egg." https://shop.wickeddevice.com/product-category/air-quality-egg; https://airqualityegg.wickeddevice.com/portal.

Williams, R., Vasu Kilaru, E. Snyder, A. Kaufman, T. Dye, A. Rutter, A. Russell, and H. Hafner. *Air Sensor Guidebook.* EPA/600/R-14/159, NTIS PB2015–100610. Washington, D.C.: US Environmental Protection Agency (2014). https://www.epa.gov/air-sensor-toolbox/how-use-air-sensors-air-sensor-guidebook.

Willow, Anna J., Rebecca Zak, Danielle Vilaplana, and David Sheeley. "The Contested Landscape of Unconventional Energy Development: A Report from Ohio's Shale Gas Country." *Journal of Environmental Studies and Sciences* 4, no. 1 (2014): 56–64.

Wolverton, B. C. *How to Grow Fresh Air: House Plants That Purify Your Home or Office.* 1996. London: Weidenfeld & Nicolson, 2008.

World Health Organization. "Air Pollution." https://www.who.int/health-topics/air-pollution.

World Health Organization. "Ambient Air Pollution: A Global Assessment of Exposure and Burden of Disease." Geneva, 2016. https://apps.who.int/iris/bitstream/handle/10665/250141/9789241511353-eng.pdf.

World Health Organization. "Ambient (Outdoor) Air Pollution." September 22, 2021. https://www.who.int/news-room/fact-sheets/detail/ambient-(outdoor)-air-quality-and-health.

World Health Organization. "Ambient (Outdoor) Air Quality and Health." Fact sheet 313, updated March 2014. http://www.who.int/mediacentre/factsheets/fs313/en.

World Health Organization. *WHO Air Quality Guidelines for Particulate Matter, Ozone, Nitrogen Dioxide and Sulfur Dioxide: Global Update 2005.* Geneva: World Health Organization, 2005.

Worthington, Andy. "Deptford's Tidemill Campaign and the Dawning Environmental Rebellion against the Dirty Housing 'Regeneration' Industry." May 24, 2019. http://www.andyworthington.co.uk/2019/05/24/deptfords-tidemill-campaign-and-the-dawning-environmental-rebellion-against-the-dirty-housing-regeneration-industry/.

Wylie, Sara Ann. *Fractivism: Corporate Bodies and Chemical Bonds.* Durham, N.C.: Duke University Press, 2018.

Wynter, Sylvia. "Unsettling the Coloniality of Being/Power/Truth/Freedom: Towards the Human, after Man, Its Overrepresentation—an Argument." *CR: The New Centennial Review* 3, no. 3 (2003): 257–337.

Wynter, Sylvia, and Katherine McKittrick. "Unparalleled Catastrophe for Our Species?" In *Sylvia Wynter: On Being Human as Praxis*, edited by Katherine McKittrick, 9–89. Durham, N.C.: Duke University Press, 2015.

Wythoff, Grant, ed. *The Perversity of Things: Hugo Gernsback on Media, Tinkering, and Scientifiction.* Minneapolis: University of Minnesota Press, 2016.

yang9741. "How to Tear Down a Digital Caliper and How Does a Digital Caliper Work." https://www.instructables.com/How-to-Tear-Down-a-Digital-Caliper-and-How-Does-a-/.

Yusoff, Kathryn. *A Billion Black Anthropocenes or None*. Minneapolis: University of Minnesota Press, 2018.

Yusoff, Kathryn, and Jennifer Gabrys. "Climate Change and the Imagination." *Wiley Interdisciplinary Review of Environmental Sciences* 2, no. 4 (2011): 516–34.

Zandbergen, Dorien. "'We Are Sensemakers': The (Anti-)Politics of Smart City Co-Creation." *Public Culture* 29, no. 3 (2017): 539–62.

Zetter, Kim. "How to Make Your Own NSA Bulk Surveillance System." *Wired*, January 27, 2016. https://www.wired.com/2016/01/how-to-make-your-own-nsa-bulk-surveillance-system.

Zhang, Yuzhong, Ritesh Gautam, Sudhanshu Pandey, Mark Omara, Joannes D. Maasakkers, Pankaj Sadavarte, David Lyon, Hannah Nesser, Melissa P. Sulprizio, Daniel J. Varon, Ruixiong Zhang, Sander Houweling, Daniel Zavala-Araiza, Ramon A. Alvarez, Alba Lorente, Steven P. Hamburg, Ilse Aben, and Daniel J. Jacob. "Quantifying Methane Emissions from the Largest Oil-Producing Basin in the United States from Space." *Science Advances* 6, no. 17 (2020): 1–9. https://doi.org/10.1126/sciadv.aaz5120.

Zitouni, Benedikte. "Planetary Destruction, Ecofeminists, and Transformative Politics in the Early 1980s." *Interface* 6, no. 2 (2014): 244–70.

Zook, Matthew, Mark Graham, Taylor Shelton, and Sean Gorman. "Volunteered Geographic Information and Crowdsourcing Disaster Relief: A Case Study of the Haitian Earthquake." *World Medical & Health Policy* 2, no. 2 (2010): 7–33.

Index

JENNIFER GABRYS is chair in Media, Culture, and Environment at the University of Cambridge. She is author of *Program Earth: Environmental Sensing Technology and the Making of a Computational Planet* (Minnesota, 2016).